Springer Theses

Recognizing Outstanding Ph.D. Research

For further volumes:
http://www.springer.com/series/8790

Aims and Scope

The series "Springer Theses" brings together a selection of the very best Ph.D. theses from around the world and across the physical sciences. Nominated and endorsed by two recognized specialists, each published volume has been selected for its scientific excellence and the high impact of its contents for the pertinent field of research. For greater accessibility to non-specialists, the published versions include an extended introduction, as well as a foreword by the student's supervisor explaining the special relevance of the work for the field. As a whole, the series will provide a valuable resource both for newcomers to the research fields described, and for other scientists seeking detailed background information on special questions. Finally, it provides an accredited documentation of the valuable contributions made by today's younger generation of scientists.

Theses are accepted into the series by invited nomination only and must fulfill all of the following criteria

- They must be written in good English.
- The topic should fall within the confines of Chemistry, Physics, Earth Sciences, Engineering and related interdisciplinary fields such as Materials, Nanoscience, Chemical Engineering, Complex Systems and Biophysics.
- The work reported in the thesis must represent a significant scientific advance.
- If the thesis includes previously published material, permission to reproduce this must be gained from the respective copyright holder.
- They must have been examined and passed during the 12 months prior to nomination.
- Each thesis should include a foreword by the supervisor outlining the significance of its content.
- The theses should have a clearly defined structure including an introduction accessible to scientists not expert in that particular field.

Yue Wang

Low Threshold Organic Semiconductor Lasers

Hybrid Optoelectronics and Applications as Explosive Sensors

Doctoral Thesis accepted by
the University of St Andrews, UK

Springer

Author
Dr. Yue Wang
School of Physics and Astronomy,
Organic Semiconductor Centre
University of St Andrews
St Andrews
Scotland, UK

Supervisors
Dr. Graham Turnbull
School of Physics and Astronomy,
Organic Semiconductor Centre
University of St Andrews
St Andrews
Scotland, UK

Prof. Ifor D. W. Samuel
School of Physics and Astronomy,
Organic Semiconductor Centre
University of St Andrews
St Andrews
Scotland, UK

ISSN 2190-5053 ISSN 2190-5061 (electronic)
ISBN 978-3-319-01266-7 ISBN 978-3-319-01267-4 (eBook)
DOI 10.1007/978-3-319-01267-4
Springer Cham Heidelberg New York Dordrecht London

Library of Congress Control Number: 2013947782

Printed on acid-free paper

Springer is part of Springer Science+Business Media (www.springer.com)

Supervisors' Foreword

Advances in technology are frequently enabled by developments in materials. An excellent example is the field of plastic electronics, which is underpinned by a class of functional materials called organic semiconductors. These materials are unusual as they combine the properties of light emission and electrical conductivity with the processing of plastics, to enable optoelectronic devices such as light-emitting diodes (LEDs) and solar cells that can be fabricated by printing. For example, their application as organic light emitting diodes (OLEDs) has led to advances in full-colour light-emitting displays, now mass produced for the smartphone market and soon to appear in televisions.

The work in Yue Wang's thesis is focused on the development of a different type of organic light source—the organic semiconductor lasers. These lasers take advantage of the simple processing and scope for integration of organic materials with other technologies. The research advances two particular areas: the development of miniature plastic lasers that can be powered by a single LED, and the application of these lasers as sensors for vapours of nitroaromatic explosives. Complementary work on star-shaped laser materials and simple nanofabrication is also presented. Most of the work presented in the thesis has been undertaken as part of the HYPIX project (a collaboration between the Universities of St Andrews, Strathclyde, Edinburgh and Imperial College, and funded by the UK Engineering and Physical Sciences Research Council) which aims to develop hybrid semiconductor arrays of smart light emitting pixels.

Organic semiconductor lasers are very compact visible lasers—many can fit in the size of a postage stamp. They are made from a thin light-emitting polymer film deposited on a corrugated substrate. When illuminated with light pulses, the polymer absorbs, re-emits and amplifies the light; the wavelength-scale corrugation reflects the amplified light repeatedly through the film and diffracts a portion out as the laser beam. The research described in Yue Wang's thesis drives forward the technology and integration of organic lasers with nitride LEDs. A notable advance is the demonstration of very low power polymer lasers fabricated by ultraviolet nanoimprinting; these reduce state-of-the-art threshold densities of nanoimprinted lasers by two orders of magnitude. This advance enables the routine use of a blue LED to power the polymer laser. The plastic lens on a commercial

LED can now be replaced with a corrugated plastic film to convert the diverging LED light into a laser beam, providing a simple route to visible lasers.

The same organic semiconductors can also provide functionality as a sensing element for detecting very low concentration vapours of explosives. A molecular-scale interaction between an explosive molecule (such as TNT) and the light-emitting polymer causes a reversible dimming of the light emission. The research in this thesis demonstrates that a very fast detection can be achieved when using polymer lasers, and that this can be further enhanced by using luminescent, microporous polymers. The sensing functionality of these materials can be integrated with other optoelectronic technologies. Yue has made a polymer laser sensor for explosives integrated on an LED. Together with collaborators at the Universities of Edinburgh and Strathclyde, she has also demonstrated a prototype CMOS-based microsystem sensor that detects the presence of explosive vapours. This sensor on a computer chip operates by detecting a change in the lifetime of the polymer fluorescence following excitation by a short light pulse from an integrated micro-LED array.

St Andrews, July 2013

Dr. Graham Turnbull
Prof. Ifor Samuel

Abstract

This thesis presents studies of organic semiconductor lasers, including their operation when pumped by a light-emitting diode (LED), and their application as explosive sensors.

The photophysics and amplified spontaneous emission (ASE) of star-shaped oligofluorene truxene molecules were investigated. These materials exhibit high gain and low optical loss in thin-film waveguides. Low ASE thresholds were achieved with the truxene T3 and T4. Second-order distributed feedback (DFB) lasers were fabricated, with pump threshold intensities below 0.5 kW/cm^2 and broad tunability of the emission.

DFB lasers were demonstrated with a novel polymer BBEHP-PPV, pumped by a pulsed commercial InGaN LED. The laser emission occurred at 533 nm for peak drive current above 15 A. The output beams and pulse-dynamics of the lasers were investigated for the first time, along with a 'double-threshold' phenomenon that was observed in this long-pulse pumping regime. BBEHP-PPV lasers based on various types of diffractive resonators were also fabricated by UV nanoimprint-lithography (NIL). By optimising the resonator design and the fabrication, and the pump-beam geometry, polymer laser thresholds of $\sim$60 W/cm^2, the lowest recorded for NIL lasers, were demonstrated, enabling them to be pumped by pulsed commercial LEDs and custom micro-LED arrays.

One promising application of organic lasers is in explosive sensing. A polymer of intrinsic microporosity (PIM-1) was used to detect nitroaromatic vapours. Rapid detection of dinitrobenzene (DNB) of low vapour pressure was achieved by monitoring the photoluminescence and laser emission during exposure. In addition, a CMOS time-resolved fluorescence lifetime microsystem with a commercial green-emitting copolymer was used as a novel, portable sensor to detect DNB vapour. An InGaN LED pumped BBEHP-PPV laser was also used as a miniature sensor to detect 10 ppb of DNB. These highly sensitive hybrid sensors could be used in humanitarian demining, complementing existing technologies leading to improvement in the detection of hazardous objects.

Acknowledgments

First and foremost I offer my sincerest gratitude to my supervisors, Dr. Graham Turnbull and Prof. Ifor Samuel, who have supported me throughout my Ph.D. with their wide knowledge and personal guidance. Without Graham's encouragement, patience and effort this thesis would not have been possible.

I wish to express my warm and sincere thanks to my colleagues in the Organic Semiconductor Optoelectronics group, especially Dr. Gordon Hedley for his valuable advice and friendly help; Neil Montgomery for being a Ph.D. mate; Dr. Arvydas Ruseckas and Dr. Ashu Bansal for always helping anytime I needed them. Also past group members, Dr. Ying Yang and Dr. Georgios Tsiminis, who essentially taught me everything I know in the labs. I also wish to express my warm and sincere thanks to the technicians at the School of Physics, George Robb, Steve Balfour, Callum Smith, Chris Booth and Mark Ross.

During this work I have closely collaborated with many colleagues at the Institute of Physics and Chemistry Department at the University of Strathclyde, and University of Edinburgh. I wish to extend my warmest thanks to all those who have helped me with my work in the HYPIX project.

I owe my loving thanks to my fiancé Dr. Christopher Reardon. Without his encouragement which sometimes meant me having to stay up overnight to work on parts of my thesis it would have been impossible for me to finish this work on time. I wish to also thank my future parents-in-law, Susan for proof reading this thesis and Paul for taking the mick of my English. Last but not the least, my special gratitude is due to my parents Xinnian Ye and Qing Wang for their continuing love and support throughout my life.

Acknowledgments

Contents

Publications Relating to this Work

Y. Wang, A. L. Kanibolotsky, P. J. Skabara, I. D. W. Samuel and G. A. Turnbull, "LED pumped polymer laser sensor for explosives," *Laser & Photonics Reviews*, accepted, **2013**.

G. Tsiminis,* Y. Wang,* A. L. Kanibolotsky, P. J. Skabara, I. D. W. Samuel and G. A. Turnbull, "A nanoimprinted organic semiconductor laser pumped by an inorganic light-emitting diode," *Adv. Mat.* **2013**, *25* (20), *2826–2830.*

* Joint lead author

Y. Wang, Y. Yang, G. A. Turnbull and I. D. W. Samuel, "Explosive Sensing Using Polymer Lasers," *Mol. Cryst. Liq. Cryst.* **2012**, *554*, 103–110.

Y. Wang, B. R. Rae, R. K. Henderson, Z. Gong, J. Mckendry, E. Gu, M. D. Dawson, G. A. Turnbull and I. D. W. Samuel, "Novel Ultra-portable Explosives Sensor Based on a CMOS Fluorescence Lifetime Analysis Micro-System," *AIP Adv.* **2011**, *1* (3), 032115.

Y. Wang, G. A. Turnbull and I. D. W. Samuel, "Conjugated polymer sensors for explosive vapour detection," *Organic Semiconductors in Sensors and Bioelectronics IV, Proc. of SPIE* **2011**, *8118*, 81180.

Y. Wang, N. B. McKeown, K. J. Msayib, G. A. Turnbull and I. D. W. Samuel, "Laser Chemosensor with Rapid Responsivity and Inherent Memory Based on a Polymer of Intrinsic Microporosity," *Sensors* **2011**, *11* (3), 2478–2487.

Y. Chen, J. Herrnsdorf, B. Guilhabert, A. L. Kanibolotsky, A. R. Mackintosh, Y. Wang, R. A. Pethrick, E. Gu, G. A. Turnbull, P. J. Skabara, I. D. W. Samuel, N. Laurand, M. D. Dawson, "Laser action in a surface-structured free-standing membrane based on a pi-conjugated polymer-composite," *Org. Electron.* **2011**, *12* (1), 62–69.

Y. Wang, G. Tsiminis, Y. Yang, A. Ruseckas, A. L. Kanibolotsky, I. F. Perepichka, P. J. Skabara, G. A. Turnbull and I. D. W. Samuel, "Broadly tunable deep blue laser based on a star-shaped oligofluorene truxene," *Synth. Met.* **2010**, *160* (13), 1397–1400.

G. Tsiminis, Y. Wang, P. E. Shaw, A. L. Kanibolotsky, I. F. Perepichka, M. D. Dawson, P. J. Skabara, G. A. Turnbull and I. D. W. Samuel, "Low-threshold organic laser based on an oligofluorene truxene with low optical losses," *Appl. Phys. Lett.* **2009**, *94* (24), 243304.

Chapter 1
Introduction

1.1 Overview

Over the past 50 years, the development of light-emitting organic semiconductors has attracted a great deal of interest. Since the discovery of electroluminescence (EL) in organic materials in the 1950s and 1960s [1–3], along with the discovery of conductive properties of doped polyacetylene in 1977 recognised with the Nobel Prize in Chemistry in the year 2000 awarded to Alan Heeger, Alan MacDiarmid and Hideki Shirakawa [4], the field of organic optoelectronics has become a major research area. Various applications have been developed including light-emitting diodes [5, 6], photovoltaic cells [7], field-effect transistors [8], optical amplifiers and lasers [9]. Solution processible organic semiconductors including small molecules and conjugated polymers enable low-cost optoelectronic devices made on large-scale substrates by simplified fabrication techniques such as ink-jet printing and roll-to-roll printing [10]. Wavelength-scaled structures can also be constructed in these materials with a number of cost-effective nano-imprint methods, for example soft lithography [11].

1.2 Organics as Light Sources

Generally speaking there are two forms of organic semiconductor light sources: electrically injected organic light-emitting diodes (OLEDs) and optically pumped organic semiconductor lasers (OSLs). A basic OLED structure consists of an emissive electroluminescent (EL) layer sandwiched between two electrodes, an anode and a cathode; see Fig. 1.1a. The electrode pair creates a flow of electrons and holes that recombine inside the emissive layer thus emitting light. Much higher quantum efficiencies for the conversion of electrical energy to light have been realized in OLEDs based on a five-layer structure, illustrated in Fig. 1.1b, which provides better control over charge balance and exciton confinement by the addition of two organic layers, an electron transport layer and a hole transport

Y. Wang, *Low Threshold Organic Semiconductor Lasers*, Springer Theses,
DOI: 10.1007/978-3-319-01267-4_1,

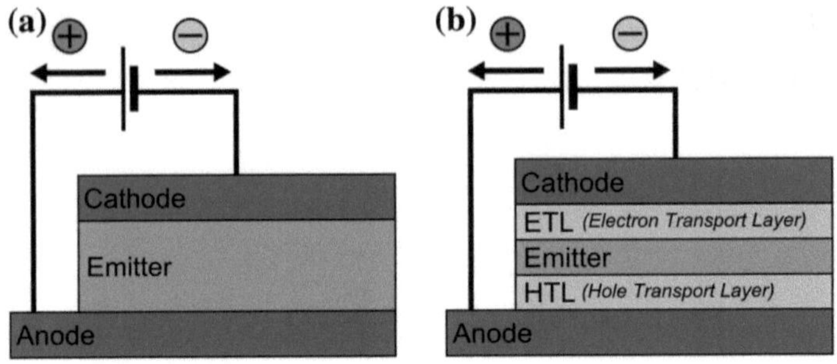

Fig. 1.1. **a** A three-layer OLED architecture; **b** a five-layer OLED with a hole-transport material (HTM) and an electron-transport material (ETM)

layer [12–14]. OLEDs with lifetimes of up to 500,000 h are now used in commercial display devices including televisions, mobile phones and cameras.

The first conjugated polymer laser was reported in 1992 by Moses [15]. It consisted of a solution of poly[2-methoxy-5-(2'-ethylhexyloxy)-p-phenylenevinylene], MEH-PPV, pumped by a frequency doubled Q-switched pulsed Nd:YAG laser. Successful laser action with defined threshold and linewidth narrowing were observed. This led the way for solid-state conjugated polymer lasers demonstrated four years later [16, 17]. It was found that lasing action could occur in simple structures containing sub-micrometer thick organic semiconductor films, indicating that the light-emitting polymers had strong absorption and high gain.

Organic semiconductors are attractive laser materials due not only to their high gain [18] but also broad spectral tunability [19, 20] and simple fabrication processing requirements [21]. Extremely low threshold lasing has been demonstrated in thin-film-based OSLs; [22, 23] the OSLs are also easy-to-handle and compact [24, 25], while keeping all the advantages of organic materials. Distributed feedback (DFB) periodic structure plays a very important role in these high performance OSLs.

Additionally, the engineering of chemical structures opens a convenient way to modify and develop optical features of a gain medium, such as absorption and emission wavelengths, Stokes shift, photoluminescence quantum yield (PLQY) and so on [26]. By choosing different materials, light emission wavelengths of OSLs can cover the whole visible spectrum [21]. With the above appealing features, organic semiconductor lasers, as coherent light sources, make themselves attractive for applications such as spectroscopy, data communications, chemosensing and lab-on-a-chip sensors for biophotonics [26–29].

Great improvements in solid-state OSLs were achieved in the last two decades, equally paced by the development of OLEDs. A question was then raised as to whether an organic laser could operate like an inorganic semiconductor laser, i.e. be pumped with an injected electrical current rather than an optical source. Although OLEDs have approached nearly 100 % internal quantum efficiency [14, 30], the highly-efficient electroluminescence (EL) is typically realized in phosphorescent emitters and at low current densities due to the charge-induced non-radiative losses increasing dramatically at high biases [31]. At the excitation densities required for electrically pumped lasing, such losses along with extra loss caused by the light absorption of the metallic contacts would make the realization of organic laser diodes particularly difficult. Substantial effort has been put into this direction; [32–36] however, no convincing demonstration of electrically pumped organic laser to date has been reported.

Another major challenge in the solid-state organic laser field is continuous wave (CW) operation. Severe issues in the progress toward this goal are losses caused by the long-lived triplet absorption that gradually builds up, reducing the gain and eventually terminating lasing action [37]. Owing to this loss mechanism, there has been no real demonstration of CW solid-state OSLs, although Bornemann and Lemmer have demonstrated a creative CW solid-state dye laser by rapidly rotating the gain medium [38]. This conventional optical pump sources for OSLs typically have pulse durations in the range of 100 fs to a few ns [9].

1.3 The Project

A novel approach to overcome the difficulty of electrically pumping OSLs is using indirect pumping techniques. Pioneering work in indirect electrically pumped OSLs has been accomplished in the Organic Semiconductor Group in St Andrews. Gallium nitride (GaN) semiconductor diode lasers were used to pump a MEH-PPV polymer laser in 2006 [24]. More recently the first light-emitting diode (LED) pumped polymer laser was demonstrated in 2008 in St Andrews [39]. This laser comprised of a corrugated fluorene-based copolymer waveguide on an indium gallium nitride (InGaN) LED, and was the closest to a directly electrically pumped OSL. It has shown a new route to inexpensive visible laser sources.

Based on the above achievements, a fully hybridized and highly compact laser source, utilising a micro-pixelated GaN inorganic semiconductor optoelectronic interface under complementary metal-oxide semiconductor (CMOS) electronic control has been proposed, and is one of the goals of the HYPIX project [40]. This PhD is part of the HYPIX project.

The generic HYPIX microsystem structure is illustrated in Fig. 1.2. Custom designed GaN micro-LEDs, driven by silicon CMOS electronics, produce blue or violet light to excite OSLs or fluorescent organic films. All these technologies can be made planar and integrated one above the other thus offering enormous prospects for a versatile optoelectronic interconnected device, which would take advantage of both inorganic and organic elements and provide pattern-programmable output to potential application interfaces such as biophotonics, visible light data communications and chemosensors.

HYPIX is a collaboration led by the University of Strathclyde, with partners including Imperial College London (ICL), University of Edinburgh and University of St Andrews. In this project, the University of Edinburgh develops the

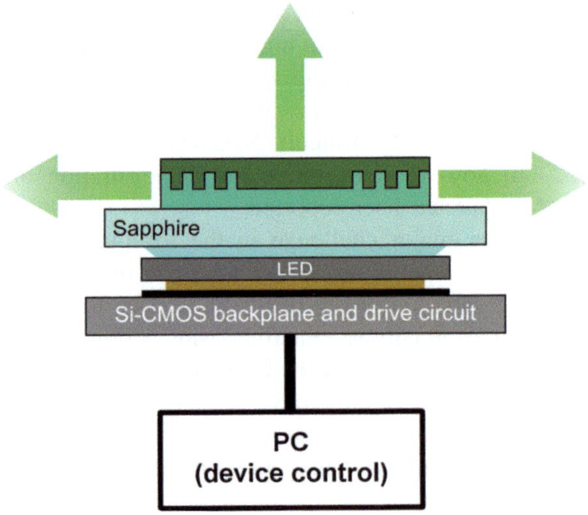

Fig. 1.2. Schematic image of the fully hybridized optoelectronic device [40]

integration of custom CMOS drive circuits; GaN micro-LED arrays are designed and fabricated at the Institute of Photonics (IoP), University of Strathclyde. St Andrews and ICL are developing the organic semiconductor lasers with gain materials provided by the Chemistry group at the University of Strathclyde. The valuable experience I have gained while working within this project is not just from the completion of ground-breaking science but also from the inter-disciplinary nature of the collaboration, working with chemists on characterising new materials and with the IoP and Edinburgh colleagues on integrating OSLs with inorganic backplanes.

1.4 The Thesis

The rest of this thesis consists of eight chapters. In Chap. 2 the theoretical background of OSLs is discussed, including both an introduction to the relevant material physics of conjugated polymers and the operation of the common laser geometries—pump sources and resonators for OSLs. The relevant methodology for their experimental study and the equipment used for material and laser characterization are outlined in Chap. 3.

Chapter 4 is the first of five experimental results chapters. A family of star-shaped truxene-core fluorene-oligomers are introduced. Key photophysical features of two long arm members, T3 and T4, including their absorption and emission spectra, and solid-state photoluminescence quantum yields are studied. To inspect their potential as gain media for lasers, optical gain and loss coefficients are measured with amplified spontaneous emission from thin film waveguides. Ellipsometry measurements are performed to examine whether the thin films are amorphous or crystalline, and to determine the optical constants, i.e. refractive indices and extinction coefficients. The refractive indices of the gain media are important parameters in designing distributed feedback lasers. Low threshold surface emitting DFB lasers, optically pumped by a Q-switched microchip laser, are demonstrated with these oligofluorene truxenes, with broad tunability in the blue spectral region.

Indirect electrical pumping of OSLs using a commercially available light-emitting diode is investigated in Chap. 5. To achieve this, low threshold organic laser materials and high power LEDs are required. A very promising gain material BBEHP-PPV was custom synthesized at the University of Strathclyde. Due to the general difficulty in the reproducing of the synthesis, the collaborators supplied various batches of the BBEHP-PPV polymer. Their photophysical and gain properties are studied in detail and the better batches are used to fabricate low threshold one-dimensional DFB lasers. A 448 nm peak emission Philips LUM-ILED GaN LED is a strong candidate for laser pumping as it has very high emitting intensity under pulsed operation. For the first time, an inorganic LED pumping an organic laser far above threshold is demonstrated in this chapter. The divergence of the output beam is measured as well as the dynamics of the polymer laser emission pulse. This pumping geometry provides a convenient and low-cost way to convert a divergent and incoherent light source into a near-collimated and coherent laser source.

Chapter 6 presents work towards the HYPIX goal of OSLs pumped by GaN micro-pixellated LEDs that could be compatible with direct CMOS control. To accommodate the lower power density provided by the micro-LEDs compared to the commercial LEDs, polymer laser thresholds require further reduction. Mixed-order DFB resonators are designed to limit the output coupling to decrease the cavity loss and thus reduce the polymer laser threshold. These gratings are initially fabricated in silicon wafers by electron-beam lithography to ensure the accuracy of the sub-hundred nanometre widths. UV nanoimprint lithography (UV-NIL) is then applied to transfer the pattern onto optically transparent substrates, i.e. glass coated with a resist material. Different pump-beam geometries are also investigated to optimize the polymer laser threshold. Very low threshold BBEHP-PPV lasers are demonstrated with the mixed-order UV-NIL gratings and successfully pumped by customized micro-LED arrays.

The compatibility of LED pumped OSLs makes them a new light source in various potential practical applications. An investigation of organic semiconductors for chemosensing nitro-aromatic compounds that are commonly used in explosives is presented in Chap. 7. A polymer of intrinsic micro-porosity (PIM–1) is of interest as its porous morphology, leading to high surface area, which can potentially enhance the penetration of explosive molecules, hence improving the sensitivity. By monitoring the photoluminescence and DFB laser emission, indication of the presence of 1,4-Dinitrobenzene (DNB) at a low vapour pressure is achieved. A significant enhancement in sensing efficiency and responsivity are established by using a DFB laser geometry. The microporous structures in PIM–1 make it a very promising material for rapid sensing of low-concentration explosives.

Chapter 8 is the last experimental chapter, and describes some potential portable explosive sensors based on organic polymers. A CMOS lifetime micro-system, developed at the University of Edinburgh, can measure nanosecond time-scale PL decay. This is adapted as a tool for trace explosive detection. The vapour molecules reversibly quench the light emission from the copolymer CDT Green and change the PL decay lifetime. The integration of the micro-LEDs as optical excitation, the CMOS micro-system as detector, and the organic semiconductor as sensing element, promises an inexpensive and ultra-portable explosive detector. At the end of the chapter, a compact LED-pumped BBEHP-PPV laser is also used for the detection of low vapour pressure explosives with very fast responsivity.

The prospect of a highly integrated and intelligent polymer laser sensor is overviewed in Chap. 9. Organic semiconductor lasers have opened up a highly sensitive way to detect trace explosives that is very important in the current state of heightened global security or for use in humanitarian demining (i.e. the clearing of landmines left after military action).

References

1. Bernanose, A., Comte, M., & Vouaux, P. (1953). A new method of light emission by certain organic compounds. *Journal of Chemical Physics, 50*, 64–68.
2. Pope, M., Kallman, H., & Magnante, P. (1963). Electroluminescence in organic crystals. *Journal of Chemical Physics, 38*, 2042–2043.
3. Dresner, J. (1969). Double injection electroluminescence in anthracene. *RCA Review, 30*, 322–334.
4. The Nobel Prize in Chemistry 2000. Retrieved from http://www.nobelprize.org/nobel_prizes/chemistry/laureates/2000/.
5. Burroughes, J. H., Bradley, D. D. C., Brown, A. R., Marks, R. N., Mackay, K., Friend, R. H., et al. (1990). Light-emitting-diodes based on conjugated polymers. *Nature, 347*(6293), 539–541.
6. Friend, R. H., Gymer, R. W., Holmes, A. B., Burroughes, J. H., Marks, R. N., Taliani, C., et al. (1999). Electroluminescence in conjugated polymers. *Nature, 397*(6715), 121–128.
7. Gunes, S., Neugebauer, H., & Sariciftci, N. S. (2007). Conjugated polymer-based organic solar cells. *Chemical Reviews, 107*(4), 1324–1338.
8. Zaumseil, J., & Sirringhaus, H. (2007). Electron and ambipolar transport in organic field-effect transistors. *Chemical Reviews, 107*(4), 1296–1323.
9. Samuel, I. D. W., & Turnbull, G. A. (2007). Organic semiconductor lasers. *Chemical Reviews, 107*(4), 1272–1295.
10. Forrest, S. R. (2004). The path to ubiquitous and low-cost organic electronic appliances on plastic. *Nature, 428*(6986), 911–918.
11. Xia, Y. N., & Whitesides, G. M. (1998). Soft lithography. *Annual Review of Materials Science, 28*, 153–184.
12. Kulkarni, A. P., Tonzola, C. J., Babel, A., & Jenekhe, S. A. (2004). Electron transport materials for organic light-emitting diodes. *Chemistry of Materials, 16*(23), 4556–4573.
13. Williams, E. L., Haavisto, K., Li, J., & Jabbour, G. E. (2007). Excimer-based white phosphorescent organic light emitting diodes with nearly 100 % internal quantum efficiency. *Advanced Materials, 19*(2), 197–202.

14. Reineke, S., Lindner, F., Schwartz, G., Seidler, N., Walzer, K., Lussem, B., et al. (2009). White organic light-emitting diodes with fluorescent tube efficiency. *Nature, 459*(7244), 234–239.
15. Moses, D. (1992). High quantum efficiency luminescence from a conducting polymer in Solution—a novel polymer laser-dye. *Applied Physics Letters, 60*(26), 3215–3216.
16. Tessler, N., Denton, G. J., & Friend, R. H. (1996). Lasing from conjugated-polymer microcavities. *Nature, 382*(6593), 695–697.
17. Hide, F., DiazGarcia, M. A., Schwartz, B. J., Andersson, M. R., Pei, Q. B., & Heeger, A. J. (1996). Semiconducting polymers: A new class of solid-state laser materials. *Science, 273*(5283), 1833–1836.
18. McGehee, M. D., & Heeger, A. J. (2000). Semiconducting (conjugated) polymers as materials for solid-state lasers. *Advanced Materials, 12*(22), 1655–1668.
19. Riechel, S., Lemmer, U., Feldmann, J., Berleb, S., Muckl, A. G., Brutting, W., et al. (2001). Very compact tunable solid-state laser utilizing a thin-film organic semiconductor. *Optics Letters, 26*(9), 593–595.
20. Heliotis, G., Xia, R., Bradley, D. D. C., Turnbull, G. A., Samuel, I. D. W., Andrew, P., et al. (2004). Two-dimensional distributed feedback lasers using a broadband, red polyfluorene gain medium. *Journal of Applied Physics, 96*(12), 6959–6965.
21. Samuel, I. D. W., & Turnbull, G. A. (2004). Polymer lasers: recent advances. *Materials Today, 7*(9), 28–35.
22. Karnutsch, C., Pflumm, C., Heliotis, G., Demello, J. C., Bradley, D. D. C., Wang, J., et al. (2007). Improved organic semiconductor lasers based on a mixed-order distributed feedback resonator design. *Applied Physics Letters, 90*, 131104.
23. Rose, A., Zhu, Z. G., Madigan, C. F., Swager, T. M., & Bulovic, V. (2005). Sensitivity gains in chemosensing by lasing action in organic polymers. *Nature, 434*(7035), 876–879.
24. Vasdekis, A. E., Tsiminis, G., Ribierre, J. C., O'Faolain, L., Krauss, T. F., Turnbull, G. A., et al. (2006). Diode pumped distributed bragg reflector lasers based on a dye-to-polymer energy transfer blend. *Optics Express, 14*(20), 9211–9216.
25. Riedl, T., Rabe, T., Johannes, H. H., Kowalsky, W., Wang, J., Weimann, T., et al. (2006). Tunable organic thin-film laser pumped by an inorganic violet diode laser. *Applied Physics Letters, 88*, 241116.
26. Grivas, C., & Pollnau, M. (2012). Organic solid-state integrated amplifiers and lasers. *Laser & Photonics Reviews, 6*(4), 419–462.
27. Chenais, S., & Forget, S. (2012). Recent advances in solid-state organic lasers. *Polymer International, 61*(3), 390–406.
28. Vannahme, C., Klinkhammer, S., Lemmer, U., & Mappes, T. (2011). Plastic lab-on-a-chip for fluorescence excitation with integrated organic semiconductor lasers. *Optics Express, 19*(9), 8179–8186.
29. Zheng, W., & He, L. (2009). Label-free, real-time multiplexed DNA detection using fluorescent conjugated polymers. *Journal of American Chemical Society, 131*(10), 3432–3433.
30. Adachi, C., Baldo, M. A., Thompson, M. E., & Forrest, S. R. (2001). Nearly 100 % internal phosphorescence efficiency in an organic light-emitting device. *Journal of Applied Physics, 90*(10), 5048–5051.
31. Baldo, M. A., Holmes, R. J., & Forrest, S. R. (2002). Prospects for electrically pumped organic lasers. *Physical Review B, 66*(3), 035321.
32. Lattante, S., Romano, F., Caricato, A. P., Martino, M., & Anni, M. (2006). Low electrode induced optical losses in organic active single layer polyfluorene waveguides with two indium tin oxide electrodes deposited by pulsed laser deposition. *Applied Physics Letters, 89*(3), 31108–31110.
33. Reufer, M., Riechel, S., Lupton, J. M., Feldmann, J., Lemmer, U., Schneider, D., et al. (2004). Low-threshold polymeric distributed feedback lasers with metallic contacts. *Applied Physics Letters, 84*(17), 3262–3264.

34. Chakaroun, M., Coens, A., Fabre, N., Gourdon, F., Solard, J., Fischer, A., et al. (2011). Optimal design of a microcavity organic laser device under electrical pumping. *Optics Express, 19*(2), 493–505.
35. Namdas, E. B., Tong, M., Ledochowitsch, P., Mednick, S. R., Yuen, J. D., Moses, D., et al. (2009). Low thresholds in polymer lasers on conductive substrates by distributed feedback nanoimprinting: Progress toward electrically pumped plastic lasers. *Advanced Materials, 21*(7), 799–802.
36. O'Neill, M., & Kelly, S. M. (2011). Ordered materials for organic electronics and photonics. *Advanced Materials, 23*(5), 566–584.
37. Zhang, Y. F., & Forrest, S. R. (2011). Existence of continuous-wave threshold for organic semiconductor lasers. *Physical Review B, 84*(24), 241301.
38. Bornemann, R., Lemmer, U., & Thiel, E. (2006). Continuous-wave solid-state dye laser. *Optics Letters, 31*(11), 1669–1671.
39. Yang, Y., Turnbull, G. A., & Samuel, I. D. W. (2008). Hybrid optoelectronics: A polymer laser pumped by a nitride light-emitting diode. *Applied Physics Letters, 92*(16), 163306.
40. HYPIX project. Retrieved from http://hypix.photonics.ac.uk/.

Chapter 2
Theory of Organic Semiconductor Lasers

2.1 Introduction

The acronym laser stands for 'light amplification by stimulated emission of radiation'. As early as 1960, Maiman demonstrated the first laser using a ruby crystal as the active medium, pumped by a flash lamp [1]. Later in 1962, the first semiconductor laser was demonstrated by two groups led by Robert Hall [2] and Marshall Nathan [3], respectively. The first visible wavelength laser diode was reported later in that same year by Holonyak [4]. These semiconductor lasers could only operate with pulsed currents and at liquid nitrogen temperature (77 K). In 1969, room-temperature, continuous-operation diode lasers using heterojunction structures were demonstrated by a team including Dmitri Garbuzov and Zhores Alferov [5] (winner of the 2000 Nobel Prize for Physics) and by the competing team of Izuyo Hayashi and Morton Panish [6]. The year 2012 marks the 50th anniversary of the invention of first semiconductor laser. During these last 50 years, semiconductor lasers have experienced remarkable development; products are now ubiquitous, covering from pure science research including spectroscopy and bio-sensing, to our daily lives such as CD/DVD players, optical mice, laser pointers and so on.

Organic semiconductor lasers (OSLs), on the other hand, are a relatively new member of the laser family. The first laser using a conjugated polymer as the gain medium was reported in 1992 by Moses [7]. In 1996, Tessler and co-workers demonstrated the first optically pumped organic microcavity laser [8]. Available OSLs nowadays are optically pumped by additional light sources such as inorganic solid-state lasers or inorganic deep-blue laser diodes, and recently light emitting diodes. With a diverse range of resonators and substrates, organic lasers offer outstanding tunability, low-cost production and mechanical flexibility.

This chapter gives an introduction to the fundamentals of organic semiconductors and lasers. In Sect. 2.2, the origins of semi-conductivity in conjugated molecules is discussed and the basic electrical and optical concepts for organic semiconductors are described. The types of commonly used organic semiconductors for lasers are listed in Sect. 2.3. The optical properties of organic

Y. Wang, *Low Threshold Organic Semiconductor Lasers*, Springer Theses,
DOI: 10.1007/978-3-319-01267-4_2,

semiconductors including absorption, fluorescence, gain and loss are described in Sect. 2.4. Section 2.5 discusses various resonators and pump sources for organic lasers as well as the output beams from these plastic lasers.

2.2 Electronic Properties

2.2.1 *Conjugated Molecules*

The essential element for an organic molecule is carbon (C). It is the building block of many types of polymers by forming long chains of interconnecting C–C bonds. Each atom of carbon consists of six electrons which occupy in different orbitals around the central nucleus. For a ground-state carbon atom, the electron configuration of $1s^22s^22p^2$ is shown in Fig. 2.1a. The two electrons in the 1s orbital are considered to be core electrons and are not available for bonding. There are four remaining valence electrons, two paired in the 2s orbital, and two unpaired in the 2p orbital which are able to form two bonds. The spatial distribution of the s and p orbitals are shown in Fig. 2.2 [9].

As energy is released when bonds form, carbon tries to maximize the number of bonds it can form. For this reason, an excited state carbon is formed by promoting one of its 2s electrons into its empty 2p orbital resulting in four unpaired electrons, which are able to form four bonds, see Fig. 2.1b. The excited state configuration is therefore $1s^22s^12p^3$. This process is favourable because of the arrangement of electrons leading to less repulsion and hence more stability.

In organic molecules, the atomic orbitals in carbon can hybridize into different geometries. The exact orbital hybridization is dependent on the bonding of the atoms. For example for a carbon atom in a methane (CH_4) molecule, sp^3 hybridization is produced. The filled 2s combines with two half filled 2p orbitals to become four half filled sp^3 hybridized orbitals. The model in Fig. 2.3a illustrates the bonding in methane. In the methane molecule, the carbon's four sp^3 hybridized orbitals point to the corners of hydrogen. The hydrogen orbitals overlap with them to form four C–H bonds which are 109.5° apart, equally spread in 3D with carbon at the centre of the tetrahedron grouping of single bonded atoms.

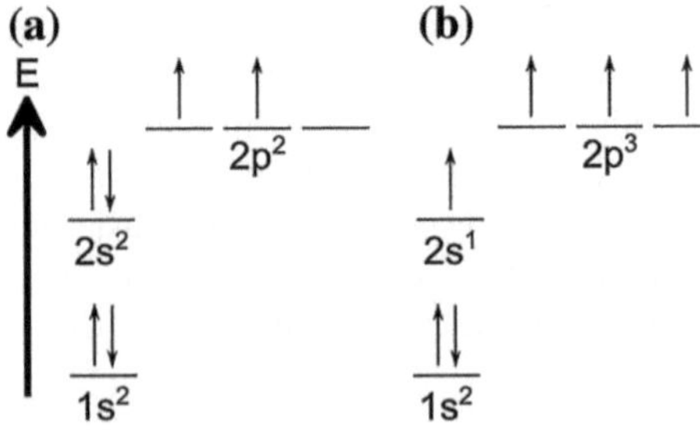

Fig. 2.1 Electron configuration of **a** ground state (lowest energy state) and **b** exited state carbon

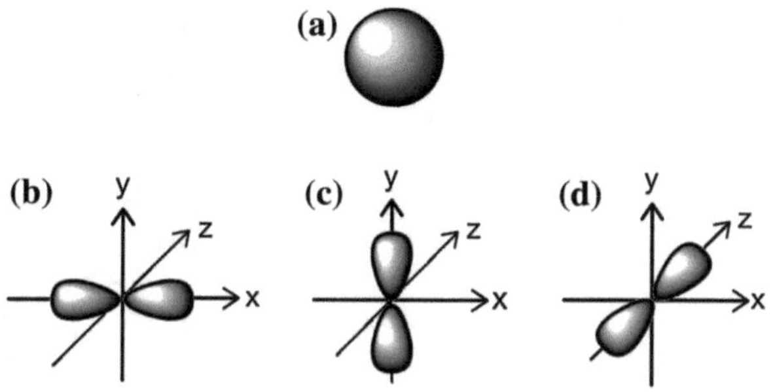

Fig. 2.2 Spatial distribution of **a** 2s and **b–d** three 2p orbital clouds

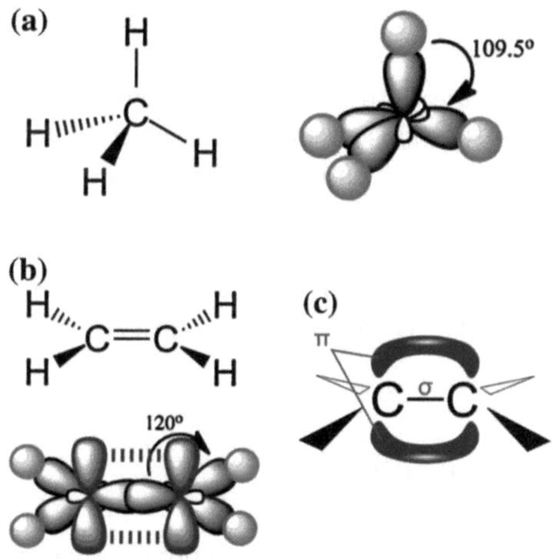

Fig. 2.3 Molecular bonding in **a** methane and **b** ethylene; **c** a C = C double bond includes both σ and π bonds

The bonding is different in trigonal planar groupings which include a C = C double bond. Considering the double bond in ethylene ($CH_2 = CH_2$), the 2s orbital and two half filled 2p orbitals mix to become three half filled sp^2 hybridized orbitals, and the fourth electron occupies an unhybridized 2p. The unhybridized 2p is perpendicular to the trigonal plane of three sp^2 hybrids which are axially symmetric and 120° apart from each other, presented in Fig. 2.3b. The two sp^2 orbitals facing each other overlap to form a σ bond, and the overlapping between the two parallel 2p orbitals is the origin of the formation of a π bond, as shown in Fig. 2.3c. In an ethylene molecule, there are four hydrogen atoms which overlap with the rest of the four sp^2 orbitals.

A backbone of conjugated polymer molecules consists of alternating single and double carbon bonds, forming a long-chain molecule, sometimes containing benzene rings. The σ bond is responsible for the rigidity of the C = C bond. More importantly, compared to the σ bond, the π bond is weak resulting in electron delocalisation along the backbone of the molecule. In other words, instead of being localised to an orbital associated with a specific bond, the π orbitals spread above and below the atoms and allow the electrons to move from one carbon to another

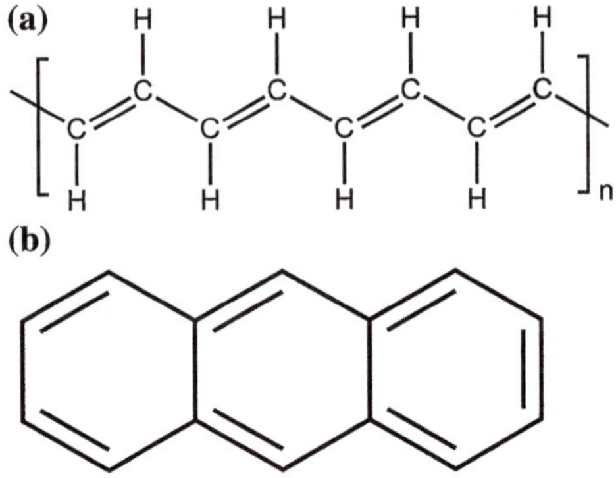

Fig. 2.4. Chemical structure of **a** polyacetylene and **b** anthracene, consisting of alternating double and single carbon bonds

along the entire backbone. Therefore directional conductivity can be produced along the axis of the so-called conjugated polymer chain. Figure 2.4 represents two typical semiconducting molecules, polyacetylene and anthracene. In an anthracene molecule, the π bond orbitals actually join up to form a ring-shaped orbital above and below the plane of each benzene ring. When three benzene rings are fused together it ends with a continuous π bond orbital region enabling conductivity, as shown in Fig. 2.5 [9].

2.2.2 Energy Levels

When two carbon atoms approach each other and form a molecular bond, it is also possible that their p-orbital electrons interfere destructively, i.e. out of phase, to

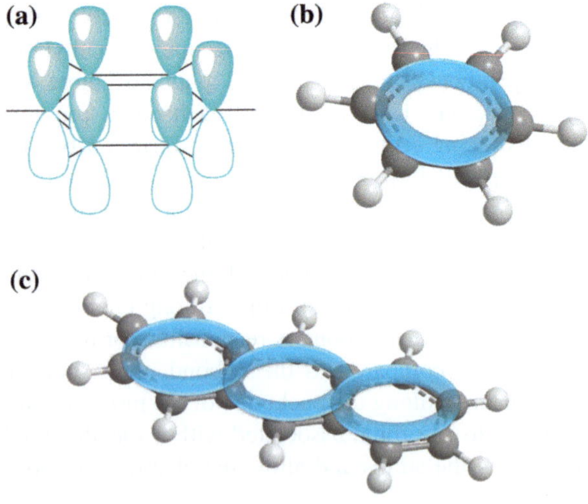

Fig. 2.5 **a** Six p-orbitals in the benzene ring; **b** these six orbitals interact with each other freely and become delocalized; **c** the π bond orbital from each ring join up in anthracene, resulting in a continuous π orbital locating above and below the molecule. The π orbitals below the molecule are not shown in this figure

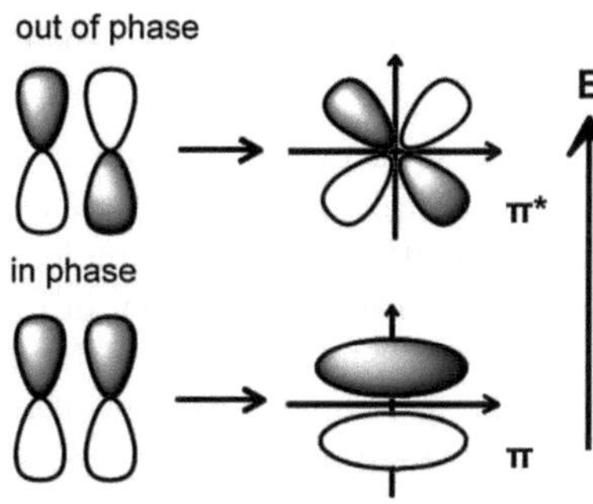

Fig. 2.6 π and π* orbitals

form an antibonding orbital π*, see Fig. 2.6. When the π* orbitals are occupied, the energy of the molecule is higher than when the π orbital is occupied. The difference in the energy of π–π* orbitals forms an energy gap. The highest energy π orbital is occupied by two electrons at thermal equilibrium and is referred to as the HOMO (highest occupied molecular orbital). The lowest unoccupied antibonding π* orbital is referred to as the LUMO (lowest unoccupied molecular orbital). The energy difference between the HOMO and LUMO is termed the HOMO–LUMO gap, i.e. energy gap. An electron can move from the π to the π* molecular orbital when there is appropriate excitation, forming an exciton (a bound electron–hole pair). Relaxation occurs with the electron moving back down to the π orbital.

Since two electrons can occupy each π or π* energy level, the spin of electrons need to be considered. There are four possible spin combinations at each energy level, such as

$$\uparrow\uparrow, \uparrow\downarrow, \downarrow\uparrow, \downarrow\downarrow \tag{2.1}$$

The singlet state is anti-symmetric hence the total spin angular momentum s = 0, while for the triplet state, s = 1.

$$|0,0\rangle = \frac{1}{\sqrt{2}}(\uparrow\downarrow - \downarrow\uparrow)\} \Rightarrow s = 0 \text{ (Singlet)} \tag{2.2}$$

$$\left.\begin{aligned} |1,1\rangle &= \uparrow\uparrow \\ |0,0\rangle &= \frac{1}{\sqrt{2}}(\uparrow\downarrow + \downarrow\uparrow) \\ |1,-1\rangle &= \downarrow\downarrow \end{aligned}\right\} \Rightarrow s = 1 \text{ (Triplet)} \tag{2.3}$$

The electron transition in singlet or triplet states is restricted by spin-selection rules [10]. Transitions are only allowed between same-spin states, i.e. singlet to singlet or triplet to triplet, whereas singlet–triplet transitions are forbidden.

2.3 Types of Organic Semiconductors

Organic semiconductors can be categorised into several groups based on their chemical structures and how they are processed. Early studies were focused on single crystals of organic molecules. A typical example is anthracene [8]; its molecular structure is shown in Fig. 2.4b, however, extremely high voltages were needed for it to emit light. The difficulty in the growing and handling of these crystals made other types of materials more attractive for light emission. Evaporation of small-molecule organic-semiconductors in order to deposit thin films that provide efficient electroluminescence attracted serious interest from the late 1980s. Aluminum tris(quinolate) (Alq_3), shown in Fig. 2.7a, is an example of small molecule organic semiconductors.

Another important member of the organic family is conjugated polymers. Development of polymerisation techniques allowed the synthesis of conjugated polymers which have long backbones of carbon, mentioned in Sect. 2.2. The most commonly used conjugated polymers as laser materials are the poly(phenylene

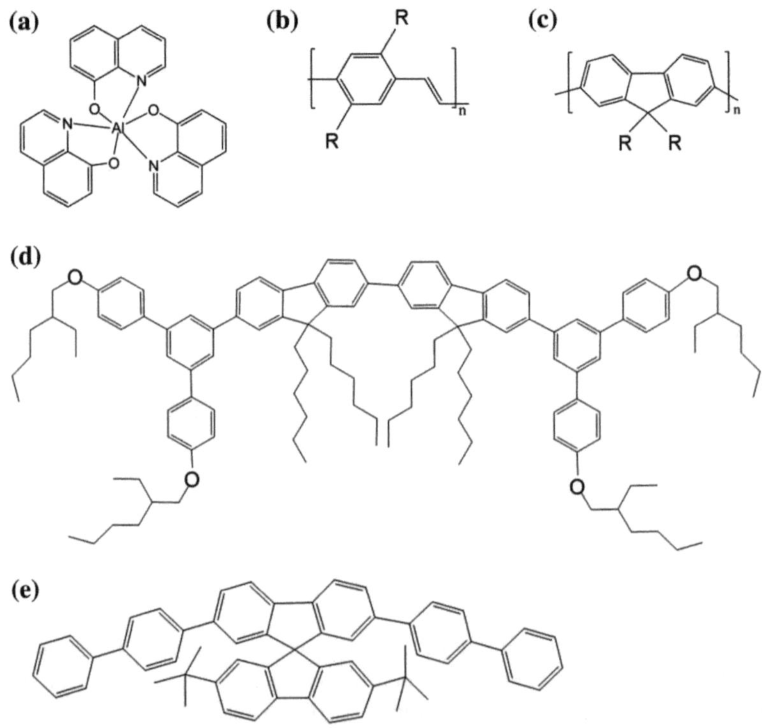

Fig. 2.7 Chemical structures of typical organic semiconductors used for lasers: **a** aluminum tris(quinolate); **b** generic poly(para-phenylene vinylene) derivative; **c** generic polyfluorene derivative; **d** bisfluorene cored dendrimer; **e** spiro-linked oligomer

vinylene) (PPV) [9, 11, 12] and the polyfluorene (PF) derivatives [13], shown in Fig. 2.7b, c respectively. A major difference from the small molecule organic semiconductors is that conjugated polymers are solution processible, which significantly simplifies the fabrication of the light-emitting devices such as lasers and light-emitting diodes. The materials discussed in Chaps. 5, 6 and 8 belong to this category.

Two further categories of organic semiconductors are conjugated dendrimers [14, 15] and oligomers [16, 17]. A conjugated dendrimer typically consists of a chromophore core defining the photophysical properties of the molecule, several conjugated dendrons (branches) controlling the intermolecular interactions of the emissive cores, and surface groups attached to their ends providing solubility. Figure 2.7d shows an example of these highly branched architectures. This is a first generation dendrimer - one level of branching with a bisfluorene core [18]. In contrast to polymers, the final category, oligomers, have a limited and known number of repeating monomer units. These materials have remarkable performance in light emitting devices [19]. Figure 2.7e shows an example of a successful lasing material, with two oligomers linked by a spiro linkage [16]. The star-shaped molecules to be discussed in Chap. 4 also belong to this category.

2.4 Optical Properties

2.4.1 Absorption and Emission of Light

As mentioned in Sect.2.2.2, the electronic transitions between π and π^* orbitals are responsible for the absorption and emission of photons. The energy of the absorbed photon equals the energy gap between π and π^* states. As the conjugation length increases, the HOMO–LUMO gap decreases, hence the energy required for the electronic transitions becomes lower, i.e. the absorption wavelength gets longer, see Table 2.1.

The interaction strength of the light with the material is expressed by the Beer Lambert Law equation:

$$I = I_0 \exp(-\alpha z) \tag{2.4}$$

where α is the absorption coefficient, z is the light path length in the material, I is the intensity of light out and I_0 is the intensity of the incident light.

The emission of light in conjugated molecules can be explained with the aid of the Franck–Condon principle [21]. As shown in Fig. 2.8, S_0 and S_1 represent the ground and first excited singlet states respectively, which split into a series of vibronic states from the lowest energy level, numbered 0 upwards. In a simplified situation, absorption occurs when an incoming photon excites an electron from the lowest electronic ground state S_0 to an unpopulated higher singlet state S_1, followed by a fast, non-radiative relaxation to the 0th vibronic level in S_1. A radiative decay process occurs from the 0th vibronic level in S_1 back to the ground state S_0

Table 2.1 The principal value of the absorption wavelength as increasing numbers of benzene rings [20]

Molecule	Structure	Absorption peak (nm)
Benzene		255 nm
Napthalene	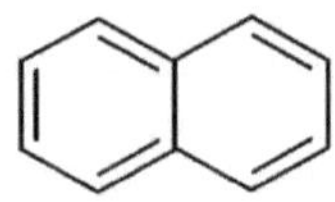	315 nm
Anthracene	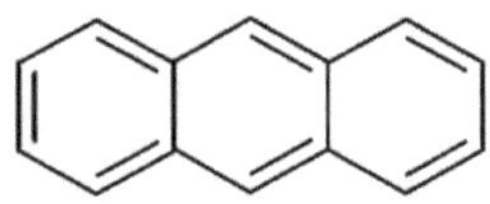	380 nm
Tetracene	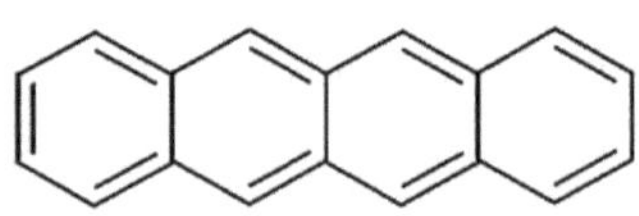	480 nm
Pentacene	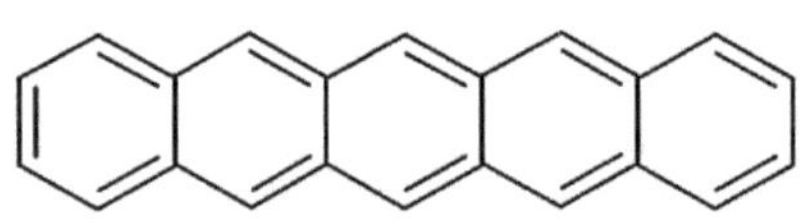	580 nm

and again, rapidly relaxes non-radiatively to the bottom of the ground state, 0th vibronic level in S_0. In conjugated molecules, the separation between the absorption and fluorescence spectrum, known as the Stokes' shift, is due to the energy loss between where a photon is absorbed and where it is emitted [22]. One common loss mechanism is the interaction between the electrons and the carbon lattice, which results in an energy displacement between the ground state and excited state under the coordinate of the distance between neighbouring carbon atoms. The disordered environment in long-chain polymers, i.e. different conjugation lengths at different locations forming various localized energy states, also extends the Stokes' shift in conjugated molecules.

A large energy shift in organic semiconductor molecules will result in less self-absorption. It also helps to increase the photoluminescence quantum yield (PLQY), leading to lower lasing thresholds. Further reduction of the spectral overlap between absorption and fluorescence spectra can be achieved by polymer blending or by doping with dye molecules [23, 24].

Optical transitions in conjugated molecules are not single wavelength transitions, but broadened as presented in Fig. 2.8. This broadening is characterised as both homogeneous and inhomogeneous broadening [25]. This means that the absorption and emission throughout the conjugated molecules have a fluctuating

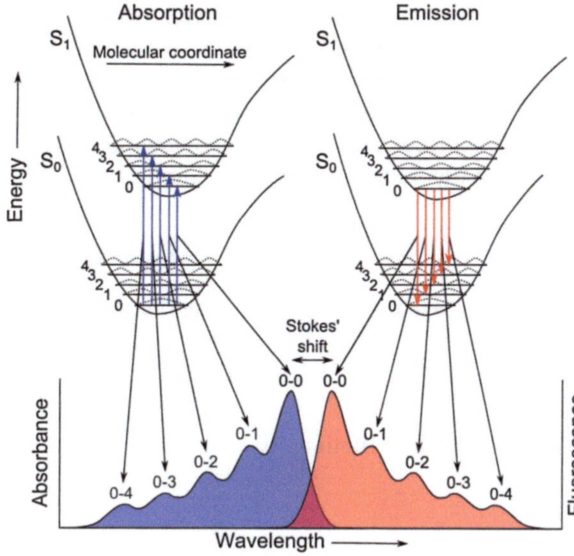

Fig. 2.8 Schematic representation of the absorption and emission spectrum with corresponding energy diagram

disorder which is the same for all the atoms, for example due to heat, and also a static disorder, for example due to conformation change such as broken conjugation in the long chains.

In a realistic situation, there are more complicated transition processes happening in a conjugated molecule, as illustrated in Fig. 2.9. When the photo-excited electron non-radiatively relaxes to the lower energy state in S_1, it is possible for the population to migrate to the triplet manifold, T1, by intersystem crossing; however this requires a spin-flip of an electron between the singlet and triplet state. To return radiatively to the ground state, another spin-flip is required. This process produces phosphorescence, however, has a much lower probability, in other words, weaker intensity and longer radiative lifetime compared to fluorescence produced with the S_1 to S_0 transition. In most organic semiconductors, phosphorescence is a very inefficient process [26]. Efficient phosphorescence were only observed in organometallic materials [27, 28].

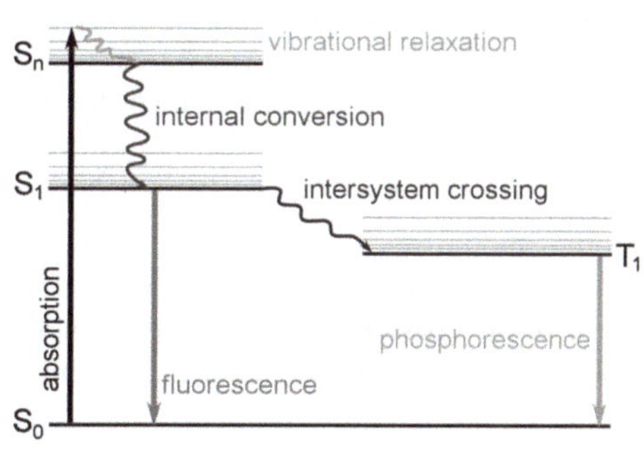

Fig. 2.9 A Jablonski diagram showing the various transition processes in a conjugated organic molecule (adapted from Ref. [20])

The efficiency of the fluorescence process is quantified by the PLQY, which is defined as the number of photons emitted divided by the number of photons absorbed [29]. With the presence of the non-radiative decay mechanisms, the PLQY of organic semiconductors is always below 100 %. The lifetime of an emissive molecule, i.e. the measured natural lifetime, τ_{PL}, is related to the radiative (k_{rad}) and non-radiative ($k_{non\text{-}rad}$) rate by Eq. (2.6):

$$\tau_{PL} = \frac{1}{k_{rad} + k_{non-rad}} \tag{2.5}$$

The fluorescence yield, Φ, is defined by the expression:

$$\Phi = \frac{k_{rad}}{k_{rad} + k_{non-rad}} \tag{2.6}$$

Thus the radiative lifetime, τ_{rad} is simply related to its natural lifetime and PLQY by:

$$\tau_{rad} = \frac{\tau_{PL}}{\Phi} \tag{2.7}$$

For triplet phosphorescence in conventional organic materials, k_{rad} is very small so that $k_{non\text{-}rad}$ dominates and the PLQY is very low.

2.4.2 Light Amplification in Organic Semiconductors

Two energy levels, E_1 and E_2 ($E_1 < E_2$), are illustrated in Fig. 2.10 to explain the phenomenon of spontaneous emission and stimulated emission. The frequency of the radiated wave is given by the well-known equation [30]:

$$\upsilon = \frac{E_2 - E_1}{h} \tag{2.8}$$

where h is Planck's constant. Spontaneous emission is characterized by the emission of a photon of energy of υh with random phase, whereas stimulated emission is induced by the incident wave and the emission from all molecules adds in phase to that of the incoming wave and in the same direction, i.e. they are coherent.

Stimulated absorption changes the upper state E_2 population, N_2 by:

$$\frac{dN_2}{dt} = B_{12}\rho_{\upsilon}N_1 \tag{2.9}$$

where B_{12} is the Einstein 'B' coefficient for the transition $E_1 \rightarrow E_2$, related to the absorption cross section, ρ is the photon density at frequency υ and N_1 is the population of the lower state E_1. In the spontaneous emission process, the rate of decay of N_2, is given by:

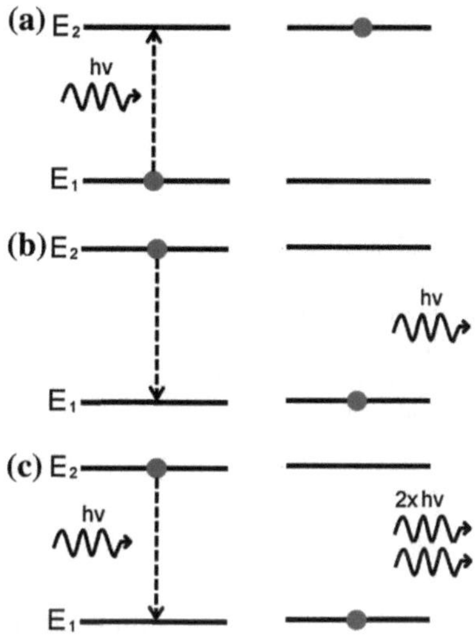

Fig. 2.10 Schematic illustration of **a** absorption, **b** spontaneous emission and **c** stimulated emission

$$\frac{dN_2}{dt} = -\frac{N_2}{\tau_2} \tag{2.10}$$

where τ_2 is the lifetime of the upper state. In the stimulated emission process, the rate of decay of N_2, is given by:

$$\frac{dN_2}{dt} = -B_{21}\rho_{\upsilon}N_2 \tag{2.11}$$

For a system in thermal equilibrium, the ratio of the population in the two energy states is given by Boltzmann statistics:

$$\frac{N_1}{N_2} = \frac{g_1}{g_2}\exp(\frac{E_2 - E_1}{kT}) = \frac{g_1}{g_2}\exp(\frac{h\upsilon}{kT}) \tag{2.12}$$

where g_1 and g_2 are the degeneracy of energy states E_1 and E_0 respectively, k is the Boltzmann's constant and T is the temperature in Kelvin.

According to Eq. (2.12), under equilibrium, $N_2 < g_2N_1/g_1$, which means the material always acts as an absorber. However, if a non-equilibrium condition is achieved, i.e. $N_2 > g_2N_1/g_1$, then the material acts as an amplifier. This condition is known as population inversion. The population inversion is the necessary condition for light amplification through a medium.

To obtain light amplification we use media with more than two energy states. As such all practical gain systems can be simplified to three or four energy levels; organic semiconductors are no exception. For a conjugated material, the energy level system resembles that of a four level laser. As the thermal relaxations are substantially faster ($\sim$10 ps) than the radiative decay ($\sim$1 ns), allowing a population inversion to be maintained relatively easily, as shown in Fig. 2.11.

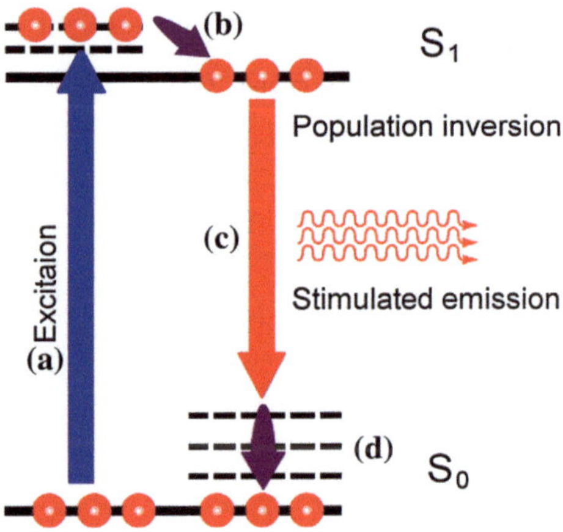

Fig. 2.11 Energy levels of the lowest two singlet states in an organic semiconductor, including the corresponding optical and thermal transitions—**a** excitation from the ground state into excited state; **b** fast non-radiative relaxation; **c** stimulated emission from S_1 to S_0; **d** fast non-radiative decay to the lowest vibronic ground state

The gain coefficient is simply the product of the stimulated emission cross section, σ, and the population inversion, defined by:

$$g = \sigma(N_2 - N_1) \tag{2.13}$$

However, several mechanisms are present in organic semiconductors that can deplete the excited state population non-radiatively. One of the mechanisms is the presence of chemical defects and impurities in a conjugated polymer, resulting in efficient quenching of the singlet photoluminescence [31]. Another possible mechanism is exciton–exciton annihilation due to collisions of excitons travelling along the backbone of a conjugated polymer [11]. This occurs at relatively high excitation densities, leading to electron and hole polaron generation thus reducing the number of the excitons that could contribute to fluorescence [32].

Apart from the above depletion mechanisms, there are some other optical losses present in the organic semiconductor gain media, including the spectrum overlap induced self-absorption that mentioned in Sect. 2.4.1, singlet excited state absorption, intersystem crossing and even triplet–triplet excited state absorption [33, 34], as illustrated in Fig. 2.12. In reality in order to achieve light amplification, sufficient gain is required to overcome not only the stimulated absorption in the material but also the various losses. With all these losses taken into consideration, the net gain

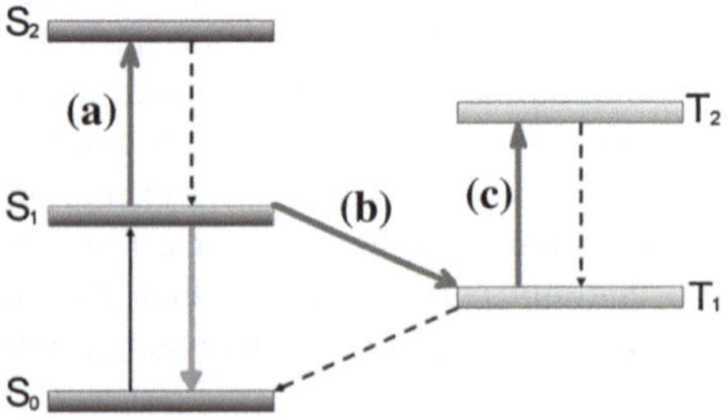

Fig. 2.12 **a** Singlet excited state absorption; **b** intersystem crossing and **c** triplet excited state absorption

(gain minus loss) in organic semiconductors can still be very high. Net gains over 60 cm^{-1} have been measured in a wide range of organic materials [35].

2.5 Organic Semiconductor Lasers

Once a population inversion is established, the gain medium will amplify an incident beam of light. To operate as a laser, another two elements are required, a feedback resonator and an excitation source. The optical amplifier is responsible for the generation of photons via either an optical or electrical excitation, while the resonator acts as a feedback cavity that forces the photons to travel several times through the amplifier and thus cause laser oscillation.

2.5.1 Resonators for Organic Lasers

The simplest organic laser resonator is two parallel mirrors with a slab of gain material placed in between, to form a Fabry–Perot cavity as shown in Fig. 2.13a. The resonant wavelengths are determined by the condition:

$$\lambda = \frac{2nL}{m} \tag{2.14}$$

where n is the refractive index of the gain material inside the cavity, L is the cavity length and m is an integer. Organic semiconductor lasers (OSLs) with this type of resonator are easy to fabricate; the organic gain medium can be either in solution or in a solid-state film [7, 13]. Due to the high gain (and strong absorption) in organic semiconductors, the gain medium thickness is required to be only in the order of hundreds of nm to achieve lasing.

Other types of resonator configurations for OSLs, including micro-ring and micro-sphere cavities, are illustrated in Fig. 2.13b, c. These types of microcavity normally lead to a multi-longitudinal mode output, and the quality of the emitted beam is usually not good as the light is emitted in multiple directions.

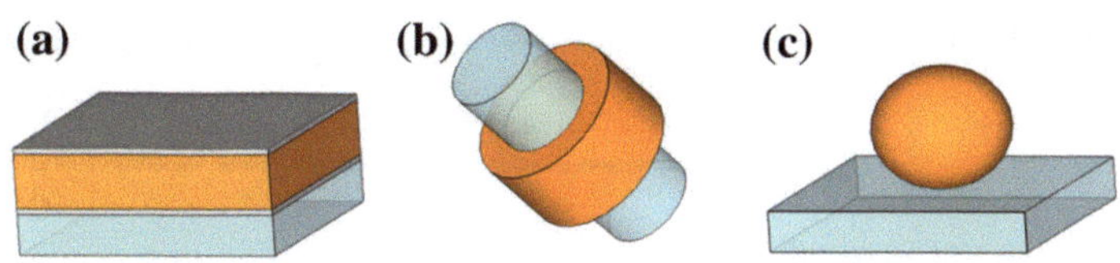

Fig. 2.13 Schematic resonators used in organic semiconductor lasers **a** Fabry–Perot microcavity; **b** micro-ring resonator, i.e. an optical fibre coated with polymer; **c** spherical microcavity

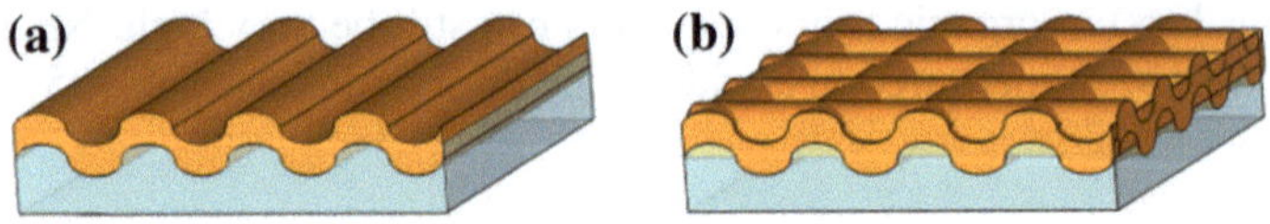

Fig. 2.14 Schematic illustration of **a** one-dimensional and **b** two-dimensional distributed feedback grating in glass coated with a layer of organic thin film

In the literature, low threshold organic semiconductor lasers have been demonstrated mainly with distributed feedback (DFB) resonators and distributed Bragg reflectors (DBR) [19, 36, 37]. In this section, I focus on the DFB cavities, which are frequently used in this thesis. Chapters 3–6 will discuss in detail the DFB cavity fabrication, their low threshold operation and the spectral tunability of emission. Figure 2.14 shows schematic of one-dimensional (1D) and two-dimensional (2D) DFB resonators.

An organic DFB laser consists of a thin film and a corrugated substrate. Waveguide theory can be used to explain how the light is confined within the organic semiconductor film [38]. Here we first do not consider the corrugation in the substrate. A slab waveguide of organic semiconductor film is structured as Fig. 2.15, allowing light to travel long distances in the planar film without diverging, i.e. the optical ray follows a repetitive, zigzag path in the film.

Guiding is achieved by using a structure consisting of a high refractive index region surrounded by regions of lower refractive indices which are known as the cladding. In this case the organic film is the high index element ($n_2 > 1.46$), sandwiched by low index air ($n_1 = 1$) and glass substrate ($n_3 = 1.46$). The light is well confined in the polymer layer by total internal reflection. The effective refractive index, n_{eff}, is determined by the whole waveguide design. The equation for waveguide mode is given as [38]

$$\frac{4\pi d}{\lambda}\sqrt{n_p^2 - n_{eff}^2} = \Phi_a + \Phi_s + 2m\pi \qquad (2.15)$$

where n_p is the refractive index of polymer, d is the polymer film thickness, λ is the light wavelength, m is an integer, ϕ_a andϕ_s are the phase shift upon total internal reflection at the polymer/air and polymer/glass interface respectively. For the transverse electric (TE) mode, ϕ_a and ϕ_s can be obtained using

Fig. 2.15 Schematic illustration of a planar optical waveguide

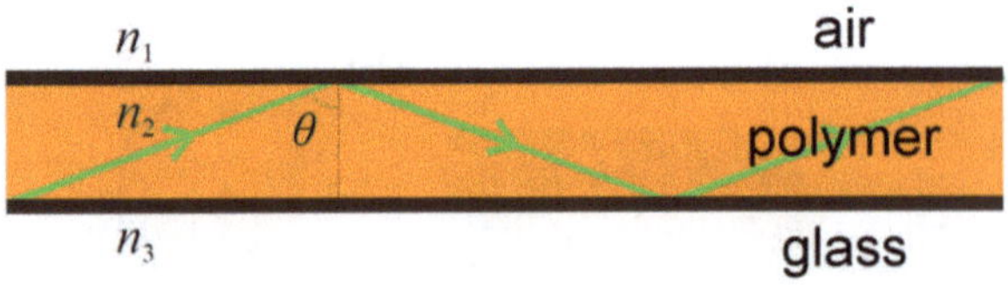

$$\phi_b = 2\ \tan^{-1}\left(\sqrt{\frac{n_{eff}^2 - n_b^2}{n_p^2 - n_{eff}^2}}\right); \quad b \equiv a, f \tag{2.16}$$

For the transverse magnetic (TM) mode, Φ_a and Φ_s can be obtained using

$$\phi_b = 2\ \tan^{-1}\left(\frac{n_p^2}{n_b^2}\sqrt{\frac{n_{eff}^2 - n_b^2}{n_p^2 - n_{eff}^2}}\right); \quad b \equiv a, f \tag{2.17}$$

The allowed range for the effective refractive index of all guided modes is defined as

$$Max\{n_a, n_s\} \leq n_{eff} \leq n_p \tag{2.18}$$

For each propagation mode there is a critical film thickness, so called cut-off thickness for example d_0 for TE_0, which means when the film is thinner than d_0 no TE mode can propagate in the film. The cut-off thickness is further discussed in Chap. 5.

Now we must take the corrugation into consideration. Instead of using mirrors, the DFB resonator provides optical feedback with a periodic corrugated microstructure. Coupled-wave theory can be used to explain how the microstructure couples free space light into and out of a waveguided mode [39, 40].

In a DFB laser, the grating leads to a periodic modulation of the effective refractive index of the optical mode propagating inside the active region. The index grating couples the forward and backward propagating waves through Bragg diffraction. Based on the coupled-wave theory, the intra-cavity electric field is written as

$$E(z) = (A(z)\exp(-ik_B z) + B(z)\exp(ik_B z))\phi(x) \tag{2.19}$$

where $\phi(x)$ describes the guided wave, A and B are the amplitudes of the forward and backward propagating waves respectively and k_B is the Bragg wave-vector, defined by:

$$k_B = \frac{2\pi n_{eff}}{\lambda_B} = \frac{m\pi}{\Lambda} \tag{2.20}$$

where m is the diffraction order, n_{eff} is the effective refractive index experienced by the guided mode, λ_B is the Bragg wavelength and Λ is the grating period. Equation (2.20) is also known as the Bragg condition for in-plane feedback in a thin film waveguide. When m = 1, the feedback is provided by first order Bragg scattering in-plane. In this case, the laser beam is emitted from the edge of the film; it is so-called edge-emitting DFB lasers or first order DFB lasers. When m = 2, it is a second order laser, whose output coupling is normal to the film plane due to the first order diffraction, and optical feedback is provided by the in-plane second order Bragg scattering, as shown in Fig. 2.16.

$A(z)$ and $B(z)$ in Eq. (2.19) are solutions of the coupled-wave equation:

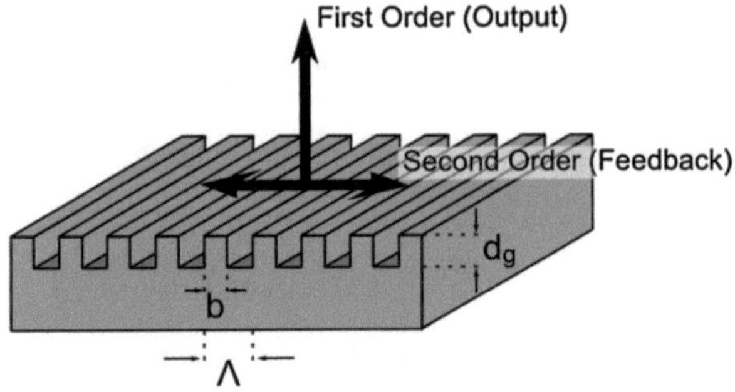

Fig. 2.16 First and second orders Bragg scattering in a surface emitting DFB laser

$$\left[\frac{\Delta\omega}{v_g} + i\left(\frac{\alpha - g}{2} + h_1\right) + i\frac{d}{dz}\right]A + (h_2 + ih_1)B = 0 \tag{2.21}$$

$$(h_2 + ih_1)A + \left[\frac{\Delta\omega}{v_g} + i\left(\frac{\alpha - g}{2} + h_1\right) - i\frac{d}{dz}\right]B = 0 \tag{2.22}$$

where h_1 is the coupling coefficient arising from two first-order diffractions and h_2 is the coupling coefficient arising from a second-order diffraction, while the other terms describe the complex changes in refractive index caused by angular frequency detuning $\Delta\omega$ from the Bragg frequency and net gain g–α, and v_g is the group velocity [40]. In a second-order DFB laser, h_1 describes the radiation coupling loss while h_2 describes the feedback coupling coefficient. The values of h_1 and h_2 are given by

$$h_1 = \frac{\omega^4 \Delta\varepsilon^2}{4k_0 k_x c^4}\left|\int dx \exp(ik_x x)\phi(x)\xi_1(x)\right|^2 \tag{2.23}$$

$$h_2 = -\frac{\omega^2 \Delta\varepsilon}{2k_0 c^2}\int dx \phi(x)^2 \xi_2(x) \tag{2.24}$$

where $\exp(ik_x x)$ approximately describes the scattered wave in the x direction, $\Delta\varepsilon$ is the dielectric change in the grating, ξ_1 and ξ_2 are the Fourier coefficients of the grating at position x. For a rectangular grating, ξ is defined by

$$\xi_m = \frac{\sin(m\pi b/\Lambda)}{m\pi} \tag{2.25}$$

where b is the width of the ridge of the grating, as shown in Fig. 2.16, b/Λ is known as the duty-cycle of the grating and m is an integer [41]. Normalized modulus of ξ_1 and ξ_1 are plotted against the duty-cycle in the range between 0 and 1 in Fig. 2.17. For a shallow grating, the relation between h_1 and h_2 can be expressed as:

$$\frac{h_1}{h_2} \cong \frac{\Delta\varepsilon d_g}{n\lambda}\tan\left(\frac{\pi b}{\Lambda}\right) \tag{2.26}$$

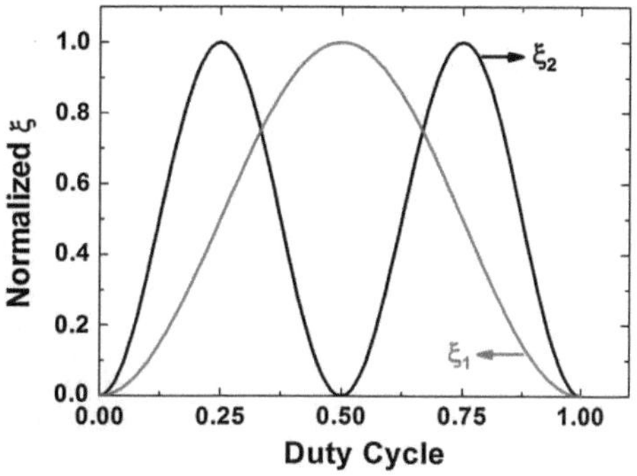

Fig. 2.17 Normalized modulus of ξ_1 and ξ_2 as a function of the duty-cycle

where d_g is the groove depth. Experimental work has shown that for 50 % duty cycle, the first-order diffraction effectively built up stimulated emission in a first-order laser hence low threshold, while a duty cycle of 25 % gives the lowest lasing threshold by optimizing the in-plane feedback in second-order DFB lasers [42]. Varying the grating depth has also shown some influence on the organic laser threshold in Navarro-Fuster *et al*'s work [43, 44].

In Eq. (2.20), it is evident that the Bragg scattering angle is wavelength dependent. In Fig. 2.18, θ is the output coupling angle of the optical mode to the normal direction and φ is the in-plane propagation direction. The parameters k_x and k_y define two wave-vector components in the plane of the waveguide, perpendicular and parallel to the grating grooves. The magnitude of the in-plane wave-vector of the guided mode, k_{mode}, is coupled into a free-space wave at angle (θ, φ) by:

$$k_{\text{mode}}^2 = \left(mk_{gx} - k_0 \sin\theta\cos\varphi\right)^2 + \left(pk_{gy} - k_0 \sin\theta\sin\varphi\right)^2 \tag{2.27}$$

where k_{gx} and k_{gy} are the two orthogonal grating vectors, while m and p determine the order of the scattering process [45]. Equation (2.27) is illustrated diagrammatically in Fig. 2.18, where the magnitude of k_{mode}, i.e. the diameter of the mode circle, is given by $2\pi n_{eff}/\lambda$, and hence varies with the photon frequency. This can be experimentally observed by an angle-resolved photoluminescence measurement in one dimension (i.e. $\varphi = 0$). Equation (2.27) can be rewritten as

$$\pm\frac{2\pi n_{eff}}{\lambda} = m\frac{2\pi}{\Lambda} - \frac{2\pi}{\lambda}\sin\theta \tag{2.28}$$

Figure 2.19 shows an example of angle-dependent emission for $\varphi = 0$ [46]. In the left figure, the 'X' shaped pattern corresponds to the PL that emits into the guided transverse mode and is subsequently Bragg scattered out of the waveguide. A is a photonic stop-band where the propagation in the waveguide is forbidden; B and C are edges of the stop-band, where standing waves form in the thin film. In the right figure, lasing occurs at point B, i.e. on the long-wavelength edge of the stop band, where the largest gain is experienced. The stop-band is seen as a dip in

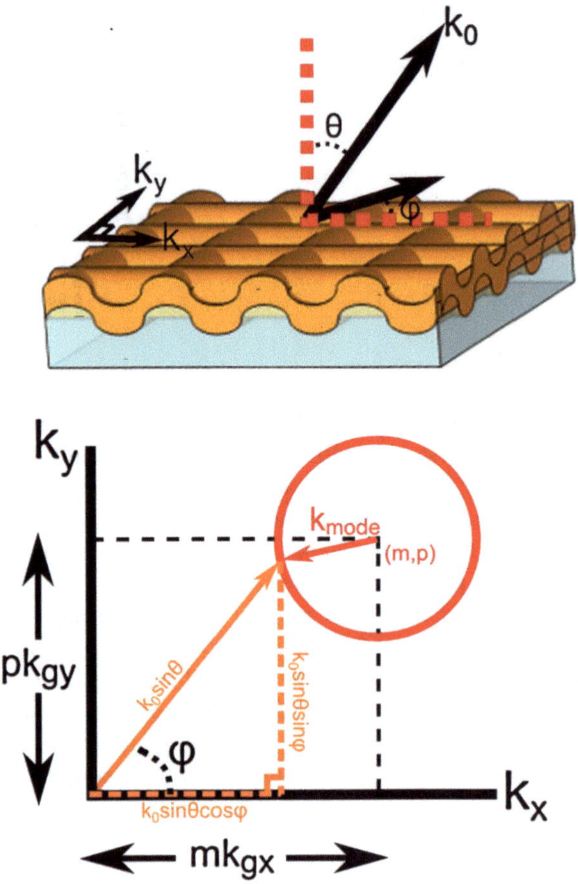

Fig. 2.18 (*Top*) light scattering in a corrugated film; (*bottom*) vector space for momentum conservation

the emission spectrum observed in the normal direction to the film below laser threshold. An example of an emission spectrum below and above laser threshold is shown in Fig. 2.20a, along with the shapes of output beams observed from a 1D and 2D second-order DFB lasers in far field. Lasing occurred only on the long-wavelength side of the stop-band because the output coupling loss of the standing wave field at the long-wavelength band edge is much lower than at the short-wavelength band edge, hence a lower lasing threshold for the long-wavelength mode [47]. The laser emission from 1D DFB resonators forms a 'double-fan' shaped pattern and is very divergent in the direction parallel to the orientation of the grating grooves due to the lack of optical confinement [48]. In the case of a 2D DFB laser, the emission pattern when the pump intensity is far above threshold is formed into a cross shape as a result of the 2D resonator, which is equivalent to two identical 1D gratings perpendicular to each other.

The out-coupled emission of the DFB laser represents a loss mechanism and thus increases the laser threshold. Therefore in theory the first-order DFB resonator

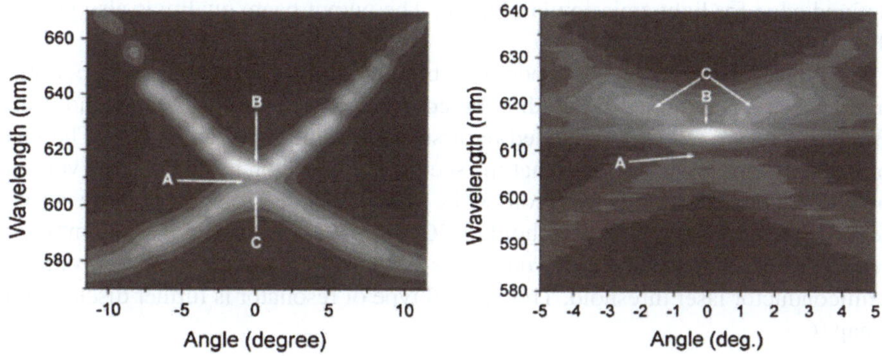

Fig. 2.19 Angle-dependent measurement for PL (*left*) and lasing (*right*) [46]. Reprinted with permission from Turnbull, G. A., Andrew, P., Jory, M. J., Barnes, W. L., Samuel, I. D. W. (2001) *Physical Review B*, *64*, 125122. Copyright (2001) by the American Physical Society

Fig. 2.20 **a** Emission spectra detected in normal direction to a MEH-PPV waveguide plane, for pump energies below and above lasing threshold [47]; output beam observed respectively from **b** a 1-D PFO laser and **c** a 2-D MEH-PPV laser

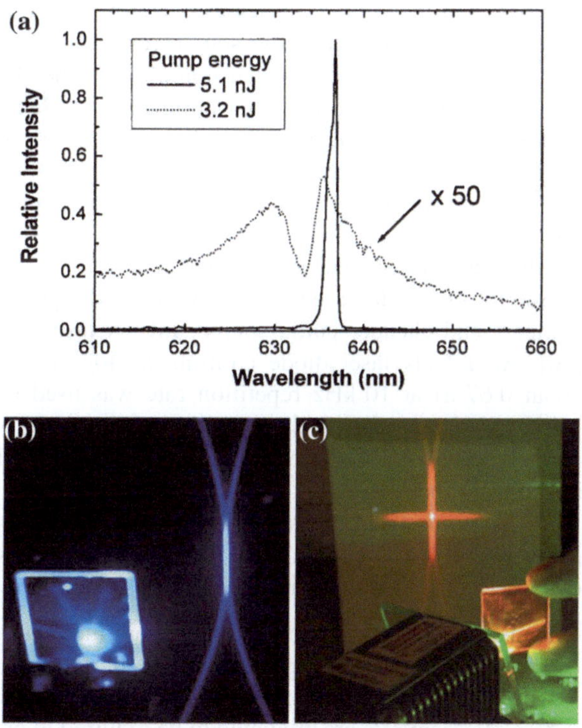

is preferred in order to achieve lower threshold operation. However, lasers based on this type of grating emit from the edge of the resonator. Due to the fact that cleaved edges are not easy with amorphous organic layers, the creation of a well defined edge for light emission is difficult. The output beam quality is also poor in this case.

In 2007, Karnutsch et al. reported a blue-emitting fluorene-based copolymer laser based on a mixed-order distributed feedback resonator - a second-order Bragg scattering region surrounded by first-order scattering regions [49]. The idea of the mixed-order resonator is that the second-order region provides some vertical outcoupling of the laser radiation while the first-order regions provide strong feedback. A very low laser threshold of 36 nJ/cm^2 (equivalent to 3.6 W/cm^2) was realized with this resonator. To date, this is the world-record for lowest organic semiconductor laser threshold. This special type of resonator is further discussed in Chap. 6.

2.5.2 Pump Sources for Organic Lasers

Compared to the simple fabrication and small size of inexpensive organic semiconductor lasers, the pumping source for organic semiconductor lasers becomes one of the major factors which complicates the design and increases the cost, which would limit the potential applications. Many different pump lasers have been used for pumping OSLs, such as dye lasers, nitrogen lasers, optical parametric oscillators (OPOs) and so on. By using a DFB resonator the laser threshold has been dramatically reduced, making pump light sources with lower output power feasible.

Microchip lasers with size about $10 \times 5 \times 5$ cm are commonly used in low threshold organic laser systems. More recently, polymer lasers pumped by a more compact and cheaper pump source, InGaN laser diode, have been demonstrated [50]. An InGaN laser diode emitting at 407 nm with maximum output energy about 0.67 nJ at 10 kHz repetition rate was used to pump DBR laser based on MEH-PPV blended with dye. The lasing threshold was as low as 0.42 nJ. Alternatively, blue-emitting lasers made of PFO doped with DPAVB were also pumped by InGaN diode lasers [51].

Since the first demonstration of an organic semiconductor laser, there has been great interest in making electrically pumped polymer lasers, however, it is extremely challenging mainly due to the following reasons [52]:

- High charge-carrier mobility in one gain material is essential in order to reduce the impact of singlet–triplet/polaron annihilations and induced absorption losses including polaron and triplet absorption. Charge-carrier mobilities of at least 5×10^{-2} cm^2/Vs are predicted to be required in order to achieve laser operation at a current density of about 10 kA/cm^2 [53], however, commonly used organic laser gain media have mobilities of only about 10^{-4} cm^2/Vs. Enhancement has been achieved, with β-phase PFO for example, the improved mobility can reach 8×10^{-3} cm^2/Vs [56];

- To pump above threshold, a current density of at least a few kA/cm^2 is required. However, the non-radiative losses such as singlet-polaron annihilation would increase with excitation strength at the same time;
- Significant losses are incurred at the metal contacts. From a study of the behaviour of Alq_3 with metal contacts, injection is seriously limited by electron trapping at the Alq_3/metal interface [54];
- Thermal stability of organic semiconductors with such high density current injection is poor [55, 56].

In 2008, a big step to address this problem was the demonstration of first indirect electrically pumped polymer laser [57]. The integration of polymer lasers with high-power nitride LEDs provided a compact and convenient electrically controlled laser device. Compared to other laser pump sources for OSLs, the InGaN LEDs are electronically and physically more robust, and much cheaper. In the same year, an electrically driven hybrid light source in the visible wavelength range was demonstrated [58]. Thin conjugated polymer films were spin-coated on nano-structured DFB gratings which were combined directly upon an inorganic LED, as shown in Fig. 2.21. With this structure, electrically generated 390 nm ultraviolet light was transferred into an organic polymer layer via optical pumping and out-coupled green light by the second-order DFB grating, although only line-narrowed fluorescence was observed in this case. Indirect electrically pumped organic semiconductor lasers are discussed in Chaps. 5 and 6.

So far all organic semiconductor lasers are pumped under pulsed conditions rather than CW conditions. The reason for this is the excitons lifetime is short—between hundreds of picoseconds and a few nanoseconds and so required high excitation densities to maintain the population inversion. Non-radiative decays from intersystem crossing, excited-state absorption and exciton–exciton annihilation can dramatically increase the cavity losses and thus reduce the net gain. The ideal pulse duration of the pump source should be less than a nanosecond with a low repetition rate so that there is enough time for the long-lived excited states to relax before the next pulse arrives. For the field of organic semiconductor lasers,

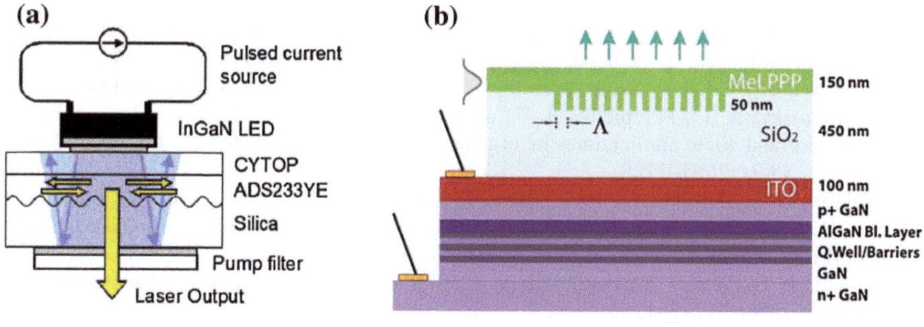

Fig. 2.21 **a** Hybrid InGaN LED pumped polymer laser; [57] **b** a compact light source that incorporates an inorganic LED, DFB grating and light-emitting polymer layer [58]

achieving CW/quasi-CW operation is an important challenge [59, 60]. The lasing thresholds with different pump pulse durations are studied in Chaps. 5 and 6.

References

1. Maiman, T. H. (1960). Optical and microwave-optical experiments in ruby. *Physical Review Letters, 4*(11), 546–566.
2. Hall, R. N., Fenner, G. E., Kingsley, J. D., Soltys, T. J., & Carlson, R. O. (1962). Coherent light emission from GaAs junctions. *Physical Review Letters, 9*(9), 366–368.
3. Nathan, M. I., Dumke, W. P., Burns, G., Dill, F. H., & Lasher, G. (1962). Stimulated emission of radiation from GaAs pn junctions. *Applied Physics Letters, 1*(3), 62–64.
4. Holonyak, Nick, & Bevacqua, S. F. (1962). Coherent (Visible) light emission from $Ga(As_{1-x}P_x)$ junctions. *Applied Physics Letters, 1*(4), 82–83.
5. Alferov, Zh I, Andreev, V. M., Portno, E. L., & Trukan, M. K. (1969). *Sov. Phys. Semicond, 3*, 1107.
6. Hayashi, I., Panish, M. B., Foy, P. W., & Sumski, S. (1970). *Applied Physics Letters, 17*, 109.
7. Moses, D. (1992). High quantum efficiency luminescence from a conducting polymer in solution—a novel polymer laser-dye. *Applied Physics Letters, 60*(26), 3215–3216.
8. Tessler, N., Denton, G. J., & Friend, R. H. (1996). Lasing from conjugated-polymer microcavities. *Nature, 382*(6593), 695–697.
9. Clayden, J., Greeves, N., Warren, S. (2001). *Organic chemistry* (2nd ed.) Oxford: Oxford University Press.
10. Lewis, G. N., & Kasha, M. (1944). Phosphorescence and the triplet state. *Journal of the American Chemical Society, 66*(12), 2100–2116.
11. Kepler, R. G., Valencia, V. S., Jacobs, S. J., & McNamara, J. J. (1996). Exciton-exciton annihilation in poly(p-phenylenevinylene) films. *Synthetic Met, 78*(3), 227–230.
12. Frolov, S. V., Vardeny, Z. V., & Yoshino, K. (1998). Cooperative and stimulated emission in poly(p-phenylene-vinylene) thin films and solutions. *Physical Review B, 57*(15), 9141–9147.
13. DiazGarcia, M. A., Hide, F., Schwartz, B. J., Andersson, M. R., Pei, Q. B., & Heeger, A. J. (1997). Plastic lasers: Semiconducting polymers as a new class of solid-state laser materials. *Synthetic Met, 84*(1–3), 455–462.
14. Lawrence, J. R., Turnbull, G. A., Samuel, I. D. W., Richards, G. J., & Burn, P. L. (2004). Optical amplification in a first-generation dendritic organic semiconductor. *Optics Letters, 29*(8), 869–871.
15. Lawrence, J. R., Namdas, E. B., Richards, G. J., Burn, P. L., & Samuel, I. D. W. (2007). Effect of generation and soft lithography on semiconducting dendrimer lasers. *Advanced Materials, 19*(19), 3000–3003.
16. Johansson, N., Salbeck, J., Bauer, J., Weissortel, F., Broms, P., Andersson, A., et al. (1998). Solid-state amplified spontaneous emission in some spiro-type molecules: A new concept for the design of solid-state lasing molecules. *Advanced Materials, 10*(14), 1136–1141.
17. Kanibolotsky, A. L., Perepichka, I. F., & Skabara, P. J. (2010). Star-shaped pi-conjugated oligomers and their applications in organic electronics and photonics. *Chemical Society Reviews, 39*(7), 2695–2728.
18. Ribierre, J. C., Tsiminis, G., Richardson, S., Turnbull, G. A., Samuel, I. D. W., Barcena, H. S., et al. (2007). Amplified spontaneous emission and lasing properties of bisfluorene-cored dendrimers. *Applied Physics Letters, 91*(8), 081108.
19. Samuel, I. D. W., & Turnbull, G. A. (2007). Organic semiconductor lasers. *Chemical Reviews, 107*(4), 1272–1295.
20. Hedley, G. J. (2010). *Ultrafast photophysics of iridium complexes*. Thesis for Doctor of Philosophy, University of St Andrews.

21. Frank, J. (1925). Elementary processes of photochemical reactions. *Transactions of the Faraday Society, 21*, 536–542.
22. Lakowicz, J. R. (1983). *Principles of fluorescence spectroscopy*. New York: Plenum Press.
23. Gupta, R., Stevenson, M., Dogariu, A., McGehee, M. D., Park, J. Y., Srdanov, V., et al. (1998). Low-threshold amplified spontaneous emission in blends of conjugated polymers. *Applied Physics Letters, 73*(24), 3492–3494.
24. Mattoussi, H., Murata, H., Merritt, C. D., Iizumi, Y., Kido, J., & Kafafi, Z. H. (1999). Photoluminescence quantum yield of pure and molecularly doped organic solid films. *Journal of Applied Physics, 86*(5), 2642–2650.
25. Pope, M., Swenberg, C. E. (1999). *Electronic processes in organic crystals and polymers* (2nd ed.). Oxford: Oxford University Press.
26. Kasha, M. (1947). Phosphorescence and the role of the triplet state in the electronic excitation of complex molecules. *Chemical Reviews, 41*(2), 401–419.
27. Baldo, M. A., O'Brien, D. F., You, Y., Shoustikov, A., Sibley, S., Thompson, M. E., et al. (1998). Highly efficient phosphorescent emission from organic electroluminescent devices. *Nature, 395*(6698), 151–154.
28. Kawamura, Y., Goushi, K., Brooks, J., Brown, J. J., Sasabe, H., & Adachi, C. (2005). 100% phosphorescence quantum efficiency of Ir(III) complexes in organic semiconductor films. *Applied Physics Letters, 86*(7), 071104.
29. Greenham, N. C., Samuel, I. D. W., Hayes, G. R., Phillips, R. T., Kessener, Y. A. R. R., Moratti, S. C., Holmes, A. B., Friend, R. H. (1995). Measurement of absolute photoluminescence quantum efficiencies in conjugated polymers. *Chemical Physics Letters 241*, 89–96.
30. Svelto, O., Hanna, D. C. (1998). *Principles of lasers* (4th ed.). New York: Springer.
31. Jenekhe, S. A., & Osaheni, J. A. (1994). Excimers and exciplexes of conjugated polymers. *Science, 265*(5173), 765–768.
32. Petrovic, J., Matavulj, P., Pinto, L., & Selmic, S. Z. (2009). Field induced singlet exciton dissociation and exciton–exciton Annihilation in MEH-PPV films studied by photocurrent spectra. *Acta Physica Polonica A, 116*(4), 595–597.
33. Shukla, A., Ghosh, H., & Mazumdar, S. (2004). Ultrafast excited state absorption and charge separation in phenylene-based conjugated polymers. *Synthetic Met, 141*(1–2), 59–65.
34. McBranch, D. W., Kraabel, B., Xu, S., Kohlman, R. S., Klimov, V. I., Bradley, D. D. C., et al. (1999). Signatures of excitons and polaron pairs in the femtosecond excited-state absorption spectra of phenylene-based conjugated polymers and oligomers. *Synthetic Met, 101*(1–3), 291–294.
35. McGehee, M. D., Gupta, R., Veenstra, S., Miller, E. K., Diaz-Garcia, M. A., & Heeger, A. J. (1998). Amplified spontaneous emission from photopumped films of a conjugated polymer. *Physical Review B, 58*(11), 7035–7039.
36. Tessler, N. (1999). Lasers based on semiconducting organic materials. *Advanced Materials, 11*(5), 363–370.
37. Chenais, S., & Forget, S. (2012). Recent advances in solid-state organic lasers. *Polymer International, 61*(3), 390–406.
38. Buckman, A. B. (1995). Guided-wave photonics. Oxford: Oxford University Press.
39. Moharam, M. G., Grann, E. B., Pommet, D. A., & Gaylord, T. K. (1995). Formulation for stable and efficient implementation of the rigorous coupled-wave analysis of binary gratings. *Journal of the Optical Society of America A: Optics, Image Science, and Vision, 12*(5), 1068–1076.
40. Kazarinov, R. F., Henry, C. H. (1985). Second-order distributed feedback lasers with mode selection provided by first-order radiation losses. *IEEE Journal of Quantum Electronics, QE-21*(2), 144–150.
41. Barlow, G. F., Shore, A., Turnbull, G. A., & Samuel, I. D. W. (2004). Design and analysis of a low-threshold polymer circular-grating distributed-feedback laser. *Journal of Optical Society of American B: Optical Physics, 21*(12), 2142–2150.

42. Turnbull, G. A., Carleton, A., Barlow, G. F., Tahraouhi, A., Krauss, T. F., Shore, K. A., et al. (2005). Influence of grating characteristics on the operation of circular-grating distributed-feedback polymer lasers. *Journal of Applied Physics, 98*(2), 023105.
43. Kogelnik, H., & Shank, C. V. (1972). Coupled-wave theory of distributed feedback lasers. *Journal of Applied Physics, 43*(5), 2327–2335.
44. Navarro-Fuster, V., Vragovic, I., Calzado, E. M., Boj, P. G., Quintana, J. A., Villalvilla, J. M., et al. (2012). Film thickness and grating depth variation in organic second-order distributed feedback lasers. *Journal of Applied Physics, 112*, 043104.
45. Turnbull, G. A., Andrew, P., Barnes, W. L., & Samuel, I. D. W. (2003). Photonic mode dispersion of a two-dimensional distributed feedback polymer laser. *Physical Review B, 67*(16), 165107.
46. Turnbull, G. A., Andrew, P., Jory, M. J., Barnes, W. L., & Samuel, I. D. W. (2001). Relationship between photonic band structure and emission characteristics of a polymer distributed feedback laser. *Physical Review B, 64*(12), 125122.
47. Turnbull, G. A., Andrew, P., Barnes, W. L., & Samuel, I. D. W. (2003). Operating characteristics of a semiconducting polymer laser pumped by a microchip laser. *Applied Physics Letters, 82*(3), 313–315.
48. Heliotis, G., Xia, R. D., Turnbull, G. A., Andrew, P., Barnes, W. L., Samuel, I. D. W., et al. (2004). Emission characteristics and performance comparison of polyfluorene lasers with one- and two-dimensional distributed feedback. *Advanced Functional Materials, 14*(1), 91–97.
49. Karnutsch, C., Pflumm, C., Heliotis, G., Demello, J. C., Bradley, D. D. C., Wang, J., et al. (2007). Improved organic semiconductor lasers based on a mixed-order distributed feedback resonator design. *Applied Physics Letters, 90*(13), 131104.
50. Vasdekis, A. E., Tsiminis, G., Ribierre, J. C., O'Faolain, L., Krauss, T. F., Turnbull, G. A., et al. (2006). Diode pumped distributed Bragg reflector lasers based on a dye-to-polymer energy transfer blend. *Optics Express, 14*(20), 9211–9216.
51. Riedl, T., Rabe, T., Johannes, H. H., Kowalsky, W., Wang, J., Weimann, T., et al. (2006). Tunable organic thin-film laser pumped by an inorganic violet diode laser. *Applied Physics Letters, 88*(24), 241116.
52. Baldo, M. A., Holmes, R. J., & Forrest, S. R. (2002). Prospects for electrically pumped organic lasers. *Physical Review B, 66*(3), 035321.
53. Gärtner, C. (2008). Organic laser diodes—modelling and simulation. Thesis for Doctor of Philosophy, University of Karlsruhe.
54. Baldo, M. A., & Forrest, S. R. (2001). Interface-limited injection in amorphous organic semiconductors. *Physical Review B, 64*(8), 085201.
55. Kulkarni, A. P., Tonzola, C. J., Babel, A., & Jenekhe, S. A. (2004). Electron transport materials for organic light-emitting diodes. *Chemistry of Materials, 16*(23), 4556–4573.
56. Prins, P., Grozema, F. C., Nehls, B. S., Farrell, T., Scherf, U., & Siebbeles, L. D. A. (2006). Enhanced charge-carrier mobility in β-phase polyfluorene. *Physical Review B, 74*, 113203.
57. Yang, Y., Turnbull, G. A., & Samuel, I. D. W. (2008). Hybrid optoelectronics: A polymer laser pumped by a nitride light-emitting diode. *Applied Physics Letters, 92*(16), 163306.
58. Butun, B., Aydin, K., Ulker, E., Cheylan, S., Badenes, G., Forster, M., et al. (2008). A hybrid light source with integrated inorganic light-emitting diode and organic polymer distributed feedback grating. *Nanotechnology, 19*, 195202.
59. Giebink, N. C., & Forrest, S. R. (2009). Temporal response of optically pumped organic semiconductor lasers and its implication for reaching threshold under electrical excitation. *Physical Review B, 79*(7), 073302.
60. Zhang, Y. F., & Forrest, S. R. (2011). Existence of continuous-wave threshold for organic semiconductor lasers. *Physical Review B, 84*(24), 241301.

Chapter 3
Experimental Methods

3.1 Introduction

In this chapter, the methodologies for the characterisation of organic semiconductor materials and fabrication of organic semiconductor lasers based on microstructured thin films are described.

Thin film sample preparation procedures are listed in Sect. 3.2 and the steady-state photophysical measurements, as well as photoluminescence (PL) decay and variable angle spectroscopic ellipsometry (VASE), that were performed for optical characterisation of various organic materials are discussed in Sect. 3.3.

As mentioned in Chap. 2, every laser consists of three basic components: the gain medium, the feedback resonator and the pump source. In Sect. 3.4, the background and methods of using amplified spontaneous emission to evaluate the gain properties of laser media are presented. In Sect. 3.5, the fabrication methods that were used to make micro-corrugated structures for the distributed feedback (DFB) lasers are discussed in detail and finally in Sect. 3.6, the optical setup and operating methodologies of optically pumped organic lasers are described, followed by a brief conclusion in Sect. 3.7.

3.2 Thin Film Preparation

The gain materials used in the fabrication of the organic semiconductor lasers in this thesis were obtained from a number of different sources; they were either purchased commercially from American Dye Source (ADS) Inc. or Cambridge Display Technology (CDT) Ltd., or synthesised by our chemistry collaborators, including Prof. Peter Skabara's group at University of Strathclyde, Prof. Neil McKeown's group at the Cardiff University and Prof. Wei Huang's group at Nanjing University of Posts and Telecommunications in China.

All the materials were supplied in powder form and stored in an oxygen and H_2O free (dry nitrogen) condition within a glove box. To prepare thin films,

Y. Wang, *Low Threshold Organic Semiconductor Lasers*, Springer Theses,
DOI: 10.1007/978-3-319-01267-4_3,

the powders were first dissolved in solution using organic solvents (spectrophotometric grade, Sigma Aldrich Co.). Depending on the molecular structures, different solvents were used to dissolve the organic materials, such as chlorobenzene, chloroform and toluene. The solutions were prepared under ambient glove box conditions in an amber vial to reduce exposure to UV light. Concentrations were typically in the range of 5–30 mg/ml. The solute was dissolved with gentle stirring, sometimes with low-temperature heating (60 °C), for a few minutes to a few hours until fully dissolved.

Good film quality is very important for achieving low-loss waveguides and low threshold lasers. To make high quality thin films (typically 10–100 s nm thick), the spin-coating method was used in this thesis. The substrate used for depositing organic materials was either glass microscope slides or fused silica/quartz glass. The substrate was first cleaned in acetone followed by isopropanol in an ultrasonic bath for 15 min each and then held on a vacuum chuck in a spin coater. Solution of organic material was spread across the substrate and then spun at a high speed, 1,000–4,000 revolutions-per-minute (rpm), for a set period of time, typically 60 s, to form a uniform film, see Fig. 3.1. The resulting film thickness is affected by several factors including spin speed, spinning time, solution concentration, substrate surface tension and solvent evaporation rate [1, 2]. For a specific solution at fixed concentration, the equation below can be used as a guideline to estimate and adjust film thickness.

$$h \propto \frac{\rho}{\sqrt{\omega}} \tag{3.1}$$

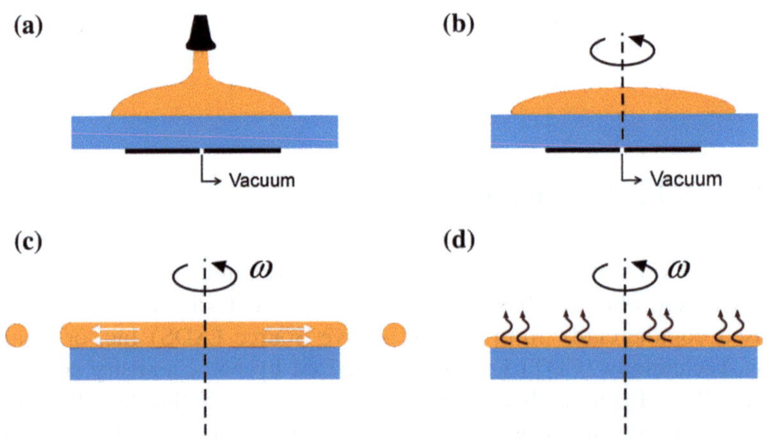

Fig. 3.1 Thin film deposition by spin-coating. **a** An excess of liquid is delivered to the substrate that is hold by vacuum underneath; **b** start spinning with an acceleration, the liquid flows outward, driven by the centrifugal force generated by the rotating substrate; **c** spinning at a constant speed, the excess solution covers the whole substrate; **d** during the spinning the solvent in the solution evaporate, resulting in a thin solid-state film of the solute [3]

where h is film thickness, ω is spin speed and ρ is concentration of the solution. An acceleration that is too high or an inappropriate spin speed or duration could result in poor film qualities. The thickness of the film was experimentally measured by a Veeco Dektak 150 surface profiler. The average film thickness was obtained with line scans across a scratch at a few different locations on the film. The accuracy for films thinner than 50 nm is low, due to the limit of the z-direction resolution and the film roughness. To measure such thin film thicknesses, other techniques were used including VASE and SEM/AFM, described in Sects. 3.3.4 and 3.5.6 respectively.

3.3 Photophysical Characterization

3.3.1 Absorption and Photoluminescence Spectroscopy

Absorption spectra were recorded with a Varian Cary 300 Bio UV–Vis spectrophotometer. Photoluminescence (PL) spectra were recorded with a Jobin Yvon FluoroMax 2 spectrometer.

Inside the Cary 300 spectrophotometer (Fig. 3.2), light was produced by an ultraviolet lamp (Deuterium) and a visible lamp (Halogen). The combination of both lamps allowed the absorption spectrum to be measured in the range of wavelength between 190 and 900 nm. The light was split into two paths, one which passed through the sample and another through a reference arm before being measured at the detector. Prior to taking a absorption spectrum, a baseline scan was performed with a clean substrate placed in each path. During the measurement, one of the clean substrates was replaced with an organic film sample and the other clean substrate remained in the other path acting as a reference. In this way, the absorption due to the substrate can be completely removed from the measured spectrum. The absorbance A of a material is defined by

$$A = -\log_{10} T \tag{3.2}$$

where T is the transmission (detected signal) with the baseline correction.

Inside the FluoroMax 2 spectrometer, light emitted from a Xenon lamp was coupled through a monochromator, allowing an excitation wavelength to be selected in the range from UV to near IR. Samples were positioned in a holder at about 45° to the excitation beam, ensuring no excitation light was reflected into the detecting direction, shown schematically in Fig. 3.3. Emission from the illuminated sample was collected through a second monochromator to a photomultiplier tube (PMT). By scanning the grating angle in the second monochromator, PL spectra were measured over a range of wavelengths. The measured spectra were corrected after the scan, for the different responsivity of the instrument at different wavelengths.

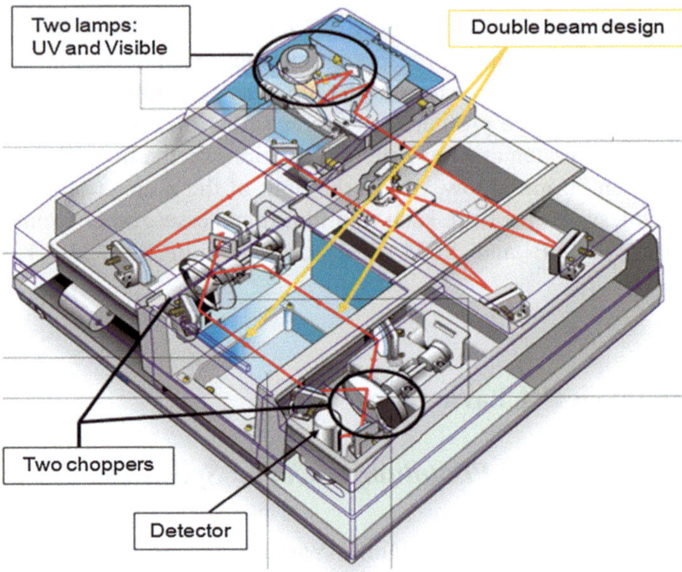

Fig. 3.2 Inside view of a Cary 300 absorption spectrometer; diagram adapted from Varian Cary 300 website [4]

Fig. 3.3 Schematic illustration of optics in Jobin Yvon FluoroMax 2 spectrometer

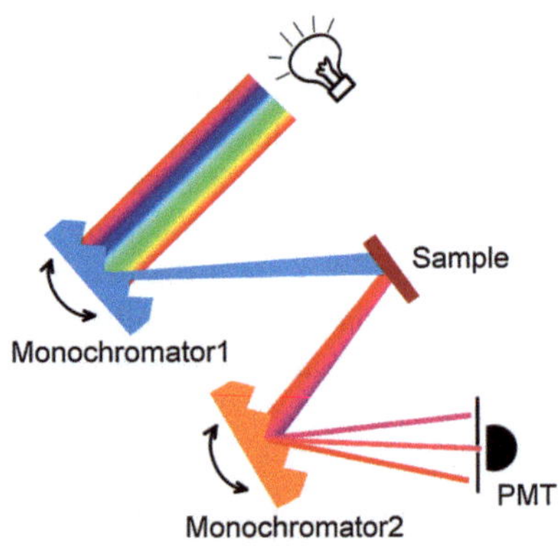

3.3.2 Photoluminescence Quantum Yield

Photoluminescence quantum yield (PLQY), a measure of how efficient a material emits light, is defined as the number of photons emitted in photoluminescence per absorbed photon. It can be measured with an integrating sphere, photodiode and an optical excitation source, with Greenham et al's method [5], schematically shown in Fig. 3.4. The film sample was mounted at the centre of the integrating sphere. The luminescence emitted in all directions was collected by the integrating sphere and measured with a calibrated silicon photodiode mounted at one of the ports of the sphere (5 in Fig. 3.4). A filter absorbing the excitation wavelength and transmitting the luminescence was inserted in front of the detector. A diffusely reflecting baffle was located near the exit port to avoid the luminescence reaching the photodiode detector directly.

A Kimmon Helium-Cadmium (He-Cd) CW laser (dual-wavelength output, 325 or 440 nm) was used in the initial experiments of my PhD. The laser entered through an entrance port (1 in Fig. 3.4) with a measured power X_{laser}. With the sample placed in the sphere but no laser incident directly on it, the sphere luminescence was recorded as X_{sphere}—the amount of luminescence due to the absorption of reflected light from the walls of the integrating sphere. With the sample excited by the laser beam, the sample luminescence, X_{sample}, was then measured. The reflectance, *R*, and transmittance, *T*, of the sample with laser power of X_{laser} respectively, were then measured with a wavelength calibrated power meter respectively. Ignoring the spectral response of the sphere, the PLQY can be given by:

$$PLQY = \frac{X_{sample} - (R + T)X_{sphere}}{(1 - R - T)X_{laser}} \tag{3.3}$$

The PLQY was then spectrally calibrated with a correction factor *y*, which was determined by the spectral response of the integrating sphere S_{sphere}, the spectral response of the photodiode $G(\lambda)$, the filter transmission spectrum, $F(\lambda)$, over the sample's emission spectrum $L(\lambda)$, calculated using:

$$y = \frac{\int S_{sphere}(\lambda)L(\lambda)G(\lambda)F(\lambda)d\lambda}{S_{sphere}(\lambda_{ex})G(\lambda_{ex})\int L(\lambda)d\lambda} \tag{3.4}$$

where λ_{ex} is the excitation wavelength. The final PLQY value was thus given by x/y [5].

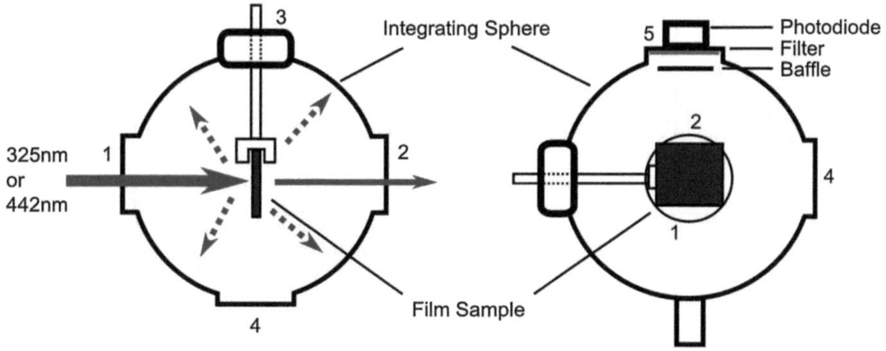

Fig. 3.4 Schematic drawing of an integrating sphere. **a** Cut-away top view; **b** cut-away front view

Solid-state PLQY in this thesis (from Chap. 5 onwards) were measured using a compact Hamamatsu Photonics C9920-02 absolute PL quantum yield measurement system, based on the principle above, see Fig. 3.5. The Hamamatsu system comprises a CW Xenon excitation light source, monochromator, and a spectrometer with an integrating sphere capable of nitrogen gas flow and a back-thinned CCD for detecting the whole spectral range simultaneously with high sensitivity [6].

3.3.3 PL Lifetime Decay

PL lifetimes were measured using a Hamamatsu streak camera. Light incident on the streak camera passed through an imaging spectrograph and so could resolve PL emission simultaneously in both time and wavelength [7].

The basic physical process underlying the operation of a streak camera is conversion of photons to photoelectrons, which are accelerated to high velocity; the photoelectrons are then deflected by a transverse electric field which rises in strength as time increases, to a phosphor screen [8]. An external trigger signal synchronizes the streak camera to the excitation pulse from a frequency doubled Titanium Sapphire oscillator, which is able to produce ultrafast pulses ($\sim$ 100 fs). A typical image from PL measurement using a streak camera is shown in Fig. 3.6. The image can be integrated over a set period of time with the brightness representing the quantity distribution of the detected photons. The position of photons from top to bottom corresponds to the time they arrived early to late, while the horizontal position from left to right corresponds to the energy of the photons low to high. This means, the time-resolved PL decay of a material, as well as its PL spectrum, can be extracted from the image. A time window of 2 ns with a resolution of 40 ps was

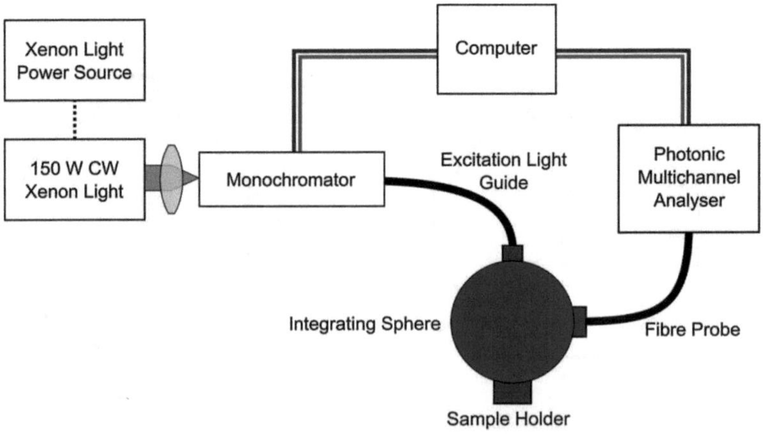

Fig. 3.5 Schematic drawing of the Hamamatsu C9920-02 PLQY measurement system

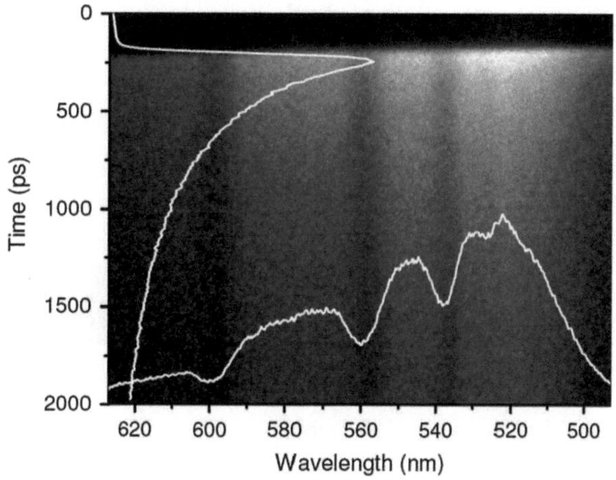

Fig. 3.6 Photoluminescence measured with the streak camera from a F8BT film; the *horizontally* and *vertically* integrated data give the PL decay and spectrum respectively; the dips in the spectrum are due to the damaged pixels on the camera [7]

used during my PhD work. In order to determine the excited-state lifetimes, fitting of the following decay function to the kinetics was required.

$$I(t) = \sum_n I_n e^{\frac{-t}{\tau_n}} \tag{3.5}$$

where *I(t)* is the intensity at the time t, I_n is a pre-exponential multiplication factor, and τ_n is the decay time. When $n = 1$, a mono-exponential decay is presented; when n > 1, it is a multi-exponential decay consisting of n individual decay components. More lifetime decay measurement methods are discussed in Chap. 8.

3.3.4 Vase

Variable angle spectroscopic ellipsometry (VASE) is a versatile and powerful technique for the investigation of the thickness, optical constants (i.e. refractive index and extinction coefficient) and optical anisotropy of thin films. It is an extremely sensitive characterization technique, non-destructive and contactless. VASE measurements in this thesis were performed using a J. A. Woolam Co. Inc. M2000-DI ellipsometer.

Ellipsometry measures the change in polarization of light upon its reflection from, or transmission through, a thin-film. A simplified diagram of an ellipsometry measurement is shown in Fig. 3.7. A sample film was mounted on a central translation stage. The incident light was produced by a white light source and linear polarized by a polarizer. Therefore, the wave travelling along the z-direction could be divided into two orthogonal components, a p-polarized component (oscillating parallel to the plane of incidence) and an s-polarized component (oscillating perpendicular to the plane of incidence and parallel to the sample surface).

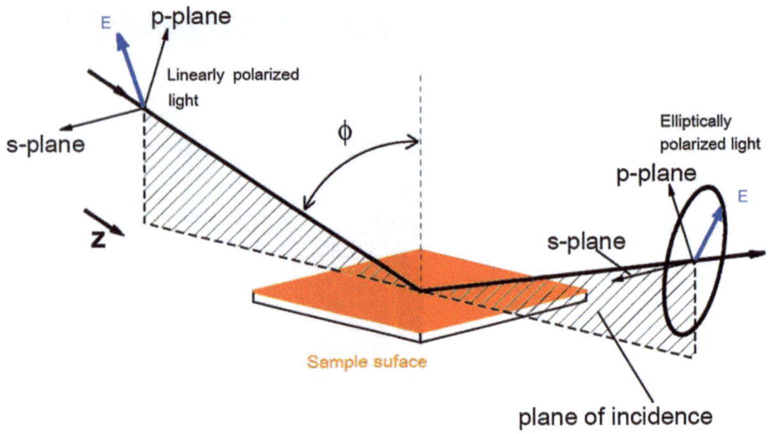

Fig. 3.7 Simplified geometry of an ellipsometry measurement, reproduced from an online tutorial http://www.jawoollam.com/tutorial_1.html

The reflected light from the sample surface, passing through a rotating polarization analyzer to a photodetector, underwent an amplitude change (Ψ) and phase shift (Δ) for both p- and s-polarized light. Ellipsometry measures the complex reflectance ratio, ρ, defined as:

$$\rho = \frac{R_p}{R_s} = \tan(\Psi)e^{i\Delta} \tag{3.6}$$

where R_p and R_s are the Fresnel reflection coefficients for the p- and s-polarized light, respectively. Very small changes in the film thickness or optical constants will alter ρ, so that ellipsometry can be used to obtain very accurate optical constants and film thicknesses, even for monolayers [9]. By measuring the relative phase and amplitude change in the reflected beam as a function of angle and optical wavelength, it was then possible to compute the optical constants of the material. The complex refractive index $\tilde{n}$ of the material is defined by

$$\tilde{n} = n + i\kappa \tag{3.7}$$

where n is the refractive index and κ is the extinction coefficient.

The standard ellipsometry measurement and data analysis procedure follows the flow chart in Fig. 3.8, using the software WVASE32. Prior to measurement the system was calibrated using a SiO_2-Si wafer with known parameters including thicknesses and optical constants. Then the experimental data from the organic semiconductor films with thicknesses in the range of 10s–100s nm were collected between 45° and 75° in intervals of 5°. The data were fitted to a model constructed based on the actual sample, see Fig. 3.9a. Initial fitting parameters such as film thickness were estimated for the preliminary calculation. The generated values were then compared to experimental data, and the fitting parameters were adjusted

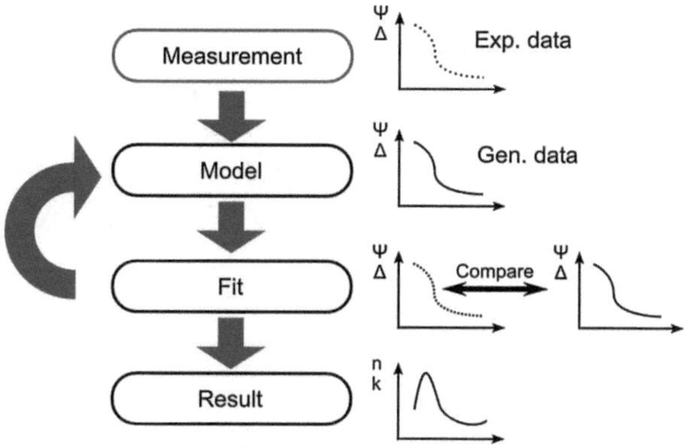

Fig. 3.8 Schematic of basic procedure for ellipsometry data analysis

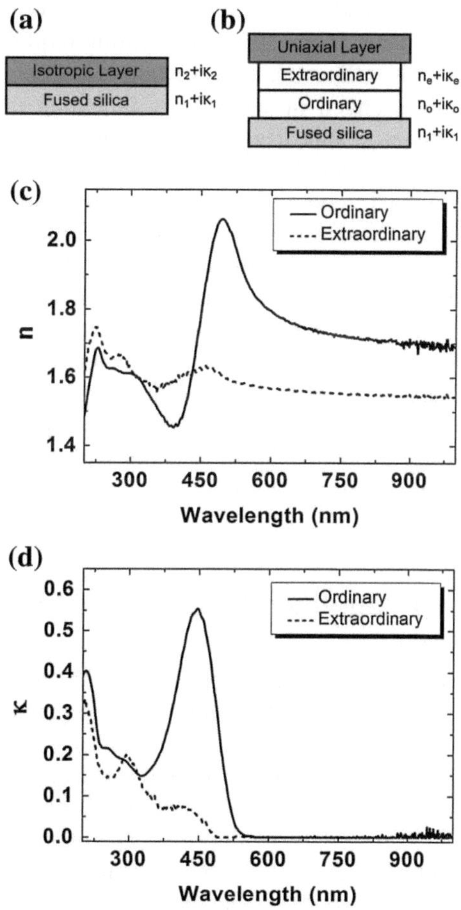

Fig. 3.9 **a** Isotropic and **b** uniaxial optical models for a polymer film on a fused silica substrate; **c** the refractive index and **d** the extinction coefficient of Super *Yellow* as derived using a uniaxial model

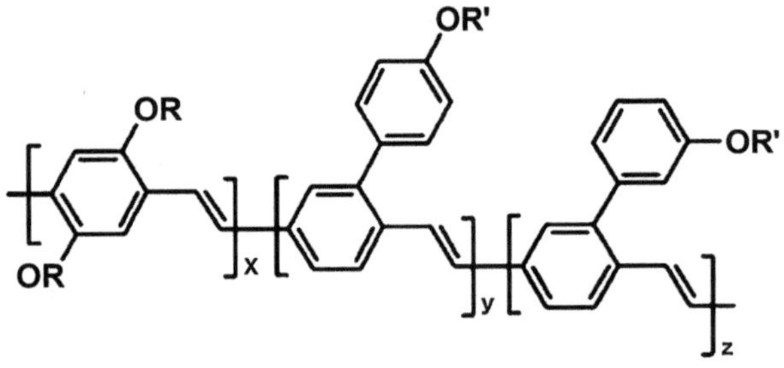

Fig. 3.10 Molecular structure of Super Yellow

iteratively to optimise the match between the experimental and calculated data graphs. A software generated Mean Squared Error (MSE) value was used to quantify the quality of the fitting. A lower MSE value corresponds to a better fitting, which leads to a more accurate determination of film thicknesses and optical constants. Typically a good fitting requires an MSE value of less than 6 [10].

The ellipsometry technique also offers insight into the way how the molecules are arranged in a thin film, including the investigation of the presence of birefringence appearing in crystalline organic molecules, i.e. anisotropic films [11]. A uniaxial model is required to analyse anisotropic films, as shown in Fig. 3.9b. In a uniaxial model, the ordinary and extraordinary optical constants are coupled within a single uniaxial layer. The thickness of the film was first determined by fitting a Cauchy model to only the transparent spectral range of the organic material, and then n and κ for ordinary and extraordinary directions were selected as fitting parameters over the full spectral range ($\sim$190–1,700 nm). The optical constants were thus obtained by a point-by-point fitting. If the optical constants are very different in the ordinary and extraordinary directions, we conclude that the film is anisotropic, otherwise isotropic. For example an efficient light-emitting material called Super Yellow with molecular structure shown in Fig. 3.10, is a highly birefringent polymer, see Fig. 3.9c and d.

3.4 Amplified Spontaneous Emission

The photophysical properties are the basic guidelines for choosing the appropriate laser materials. An ideal laser gain material should have high optical gain and low waveguiding loss to achieve low pumping thresholds. Measurement of amplified spontaneous emission (ASE) is a useful technique to evaluate the potential net gain of an organic semiconductor. To measure the ASE, an organic thin film is deposited onto a low-refractive-index substrate to form a waveguide and then optically excited by an external light source, for example an *optical parametric oscillator* (OPO). The emission is measured from the edge of the waveguide as a function of the excitation density.

3.4.1 ASE Threshold and Linewidth

Spontaneous emission happens whenever the film is optically excited, as discussed in Chap. 2. When the pump intensity is increased to a higher level, the spontaneously emitted photons can be exponentially amplified by stimulated emission as they travel through the waveguide, and finally be emitted from the end of the waveguide, as shown in Fig. 3.11. ASE is built up in this way. To distinguish ASE and laser, in ASE there is no feedback, i.e. no resonator, however, spectral linewidth narrowing can be observed in both cases.

The experimental setup for measuring the ASE threshold and full-width-at-half-maximum (FWHM) of the emission spectrum linewidth is shown in Fig. 3.12. The excitation light in ASE measurements in this thesis was generated by a tuneable OPO (Panther EX, Continuum Inc.). The third harmonic of an Nd:YAG laser (355 nm) was used to pump the OPO. The output wavelength of the OPO can be tuned from deep UV to infrared with a Beta Barium Borate (BBO) nonlinear crystal and a pair of frequency-doubling crystals, with high output power density. The OPO worked at 20 Hz repetition rate with a pulse duration of 4 ns (FWHM). Neutral density filters were placed in the beam path to adjust the intensity of the laser beam. A cylindrical lens was used to shape the excitation beam into a narrow stripe with a width of approximately 300 μm and a length which could be varied using a slit aperture. The stripe configuration was chosen because we would like to analyze ASE in a one-dimensional gain region. When ASE occurs in a long and narrow stripe, most of the light is emitted from the end of the stripe rather than all the directions [12].

For this reason, the stripe was positioned at the edge of the waveguide and the light emitted from the edge was collected and recorded by a fibre-coupled CCD spectrometer. The polymer film was placed in a vacuum chamber and pumped to approximately 10^{-4} mbar, to protect the sample from oxidation. The intensity of

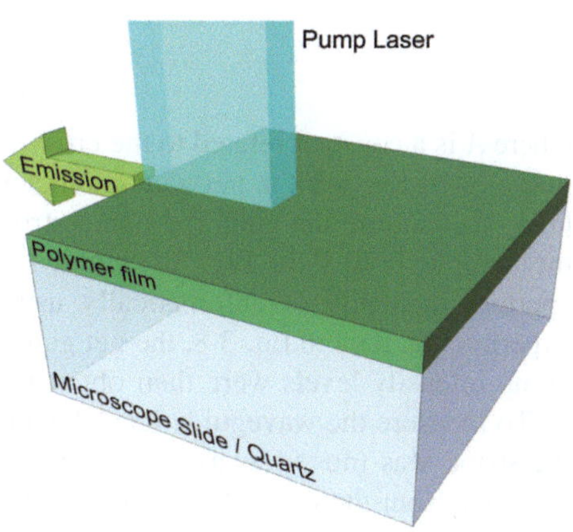

Fig. 3.11 Amplified spontaneous emission from an air-polymer-glass waveguide

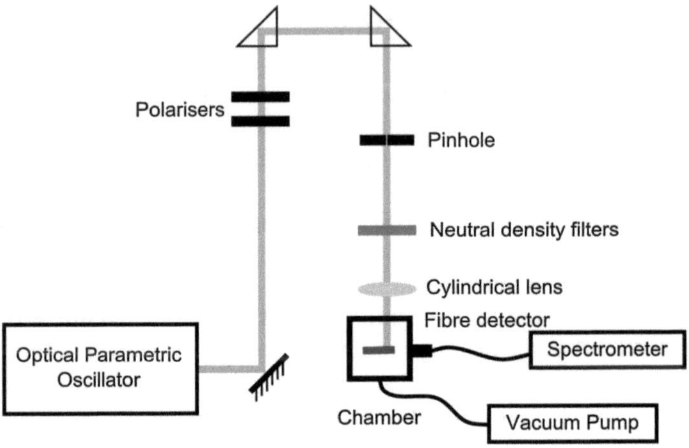

Fig. 3.12 Experimental set-up for ASE measurements

the edge-emission spectrum was plotted against the pump intensity to identify the threshold of the ASE.

Above the ASE threshold, the spontaneous emission was amplified in the waveguide structure. The wavelength of the spontaneous emission that experienced the largest gain was preferentially amplified. This caused the spectrum to be dramatically narrowed as the photons travel along the waveguide. Thus, the investigation of the dependence of the linewidth of the emission spectra on the pump intensity was another strong indication of the ASE threshold.

3.4.2 Gain and Loss Coefficients

From the theory, the output intensity from one end of the waveguide is given by:

$$I(\lambda) = \frac{A(\lambda)I_p}{g(\lambda)}(e^{g(\lambda)l} - 1) \quad (3.8)$$

where A is a constant related to the cross section for spontaneous emission, I_p is the pump intensity, g is the net gain coefficient, and l is the length of the pump stripe [12]. To measure the gain, the pump stripe was positioned right up to the edge of the sample ($x = 0$ in Fig. 3.13) and l was varied either by adjusting the slit aperture manually or automatically using a LabVIEW program. By fitting the experimental data to Eq. 3.8, the net gain coefficients of the waveguide at different pump intensity levels were then obtained.

To measure the waveguide loss, the length of the stripe l was kept constant and the stripe was moved away from the edge of the sample ($x > 0$). The emission intensity from the edge, I, is given by the expression:

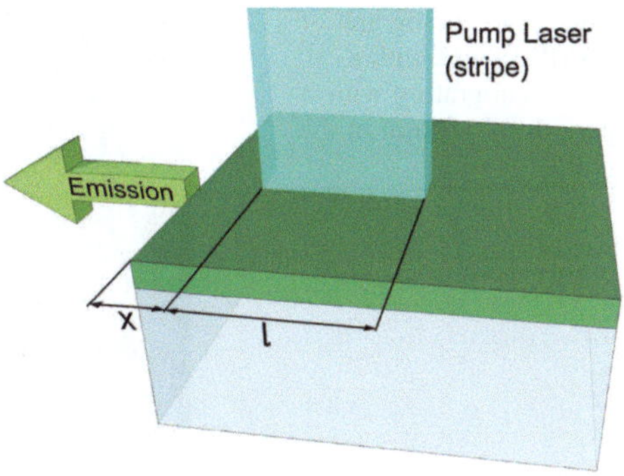

Fig. 3.13 Simplified sketch of the gain and loss measurements

$$I = I_0 \exp(-\alpha x) \tag{3.9}$$

where I_0 is the constant pump intensity, x is the length of the un-pumped region between the end of the pump stripe and the sample edge and α is the waveguide loss coefficient. Waveguide loss can arise from either self-absorption, as discussed in Chap. 2, or scattering which depends on the film quality.

The optical gain and loss coefficients of an organic material can be used to evaluate its potential performance as a laser gain medium. The combination of high gain and low loss is ideal for low-threshold lasers, at the same time, an appropriate resonator is also required.

3.5 Distributed Feedback Resonator Fabrication

In my PhD, I have used four methods to fabricate micro-resonators, including solvent-assisted micro-molding, two-beam interference holography, electron beam (e-beam) lithography, and UV nanoimprint lithography.

3.5.1 Solvent-Assisted Micro-molding

Solvent-assisted micro-molding (SAMIM) is probably the easiest way to modify the surface topography of a semiconducting polymer. The key element of SAMIM is wetting of an elastomeric mold with an organic solvent and conformal contact between the elastomeric mold and the polymer film [13].

The elastomeric mold was fabricated by blending two co-polymers, methylhydrosiloxane-dimethylsiloxane (HMS) and trimethylsiloxy-terminated vinylmethylsiloxane-dimethylsiloxane (VDT) in the ratio of 3.4 parts VDT to 1 part HMS, on top of a rigid micro-corrugated master structure. The master was placed facing up in a plastic container before the co-polymer mixture was poured over the

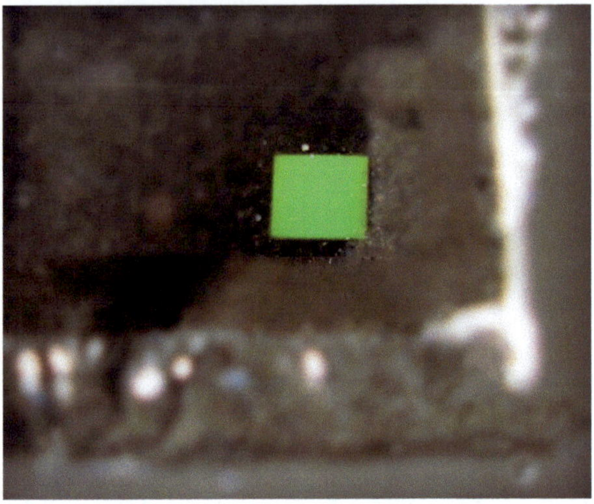

Fig. 3.14 Photograph of a SAMIM mold with a 1 × 1 cm grating with periods of 350 nm in the middle

surface. The sample was then baked at 70 °C for roughly 10 h, before the master was separated from the mold. The completed mold, see Fig. 3.14 as an example, was cooled down to room temperature before use.

The copolymer mold was resistant to most organic solvents including chlorobenzene, toluene and hexane. An appropriate solvent was used for the specific organic semiconductor in the pattern transfer process. The solvent filled the grating grooves, and then the polymer film was placed facing down onto the elastomeric mold. The mold was in contact with the polymer film for approximately 1–2 min, depending on the solvent's evaporation speed. Appropriate pressure was applied during the pattern transfer. The polymer sample was then removed from the mold, and micro-corrugations were evident on its top surface, as shown in Fig. 3.15.

With the SAMIM technique, the final quality of the pattern was directly related to the grating quality in the mold and the master. The rigid master structure can be fabricated in advance with the following techniques in Sects. 3.5.2 and 3.5.3.

3.5.2 Two-Beam Interference Holography

A two-beam interference holography setup, shown in Fig. 3.16, was built up for 1D and 2D grating fabrications. The most commonly used grating substrate was quartz plates, of 1 by 1 cm or 2.5 by 2.5 cm dimension.

In the above experimental configuration, the light source was a 325 nm He-Cd laser (the same source as used in the PLQY measurement). A beam expander, consisting of two lens, was inserted to create a relatively large diameter, collimated and uniform beam, leading to large area grating fabrication. The interferometer consisted of four UV-transparent prisms, a cube beam splitter (50:50 for

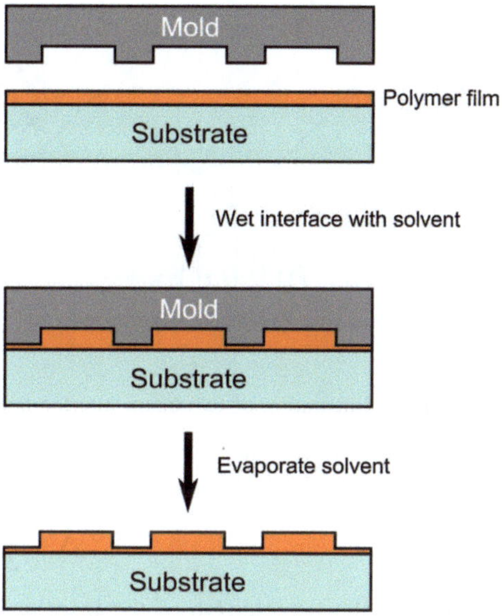

Fig. 3.15 Schematic drawing of solvent-assisted micro-molding

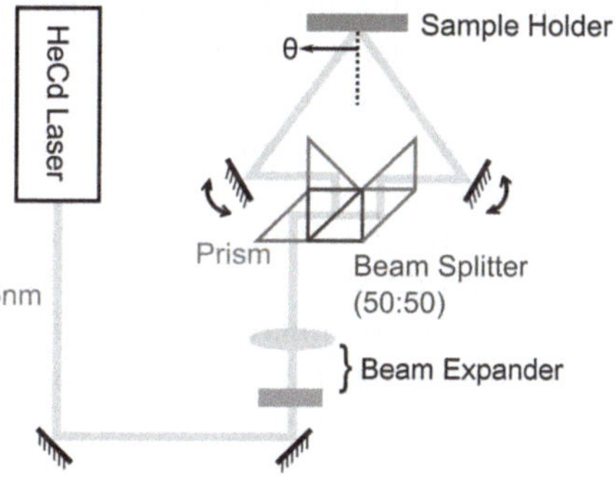

Fig. 3.16 Schematic diagram of two-beam interference holography setup

325 nm), UV-enhanced dichroic mirrors (each of them was mounted on a 360° rotation stage) and a black-anodised aluminium sample holder with a central part which could be rotated by 90° about the normal to the exposure beam for making 2D holography. The interferometer was built on a separate optical platform to prevent it from environmental vibrations. Before holographic exposure, a thin layer of diluted Microposit S1805 G2 (Chestech Ltd), a positive UV photoresist, was spin-coated onto a quartz substrate and mounted in the sample holder.

Based on optical interference theory, when the sample was exposed to two coherent UV beams incident from different directions, a standing-wave interference pattern was formed on the sample, as shown in Fig. 3.17a, and the period of the interference pattern Λ is given by:

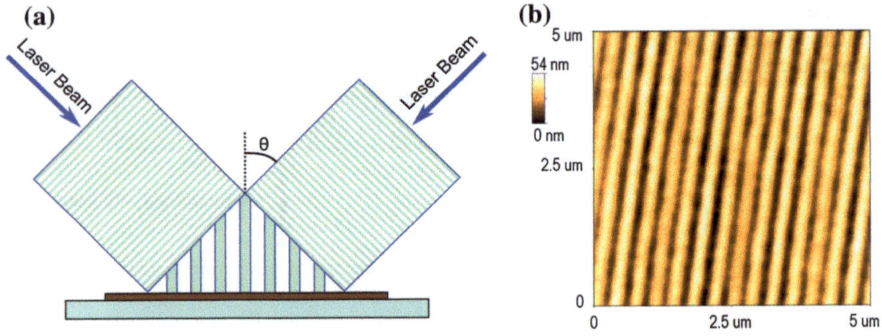

Fig. 3.17 **a** Illustration of two-beam interference; **b** an AFM image of a 1D grating made with holography

$$\Lambda = \frac{\lambda}{2\sin\theta} \tag{3.10}$$

where λ is the wavelength of the laser and θ is the angle of incidence. The period of the interference pattern can therefore be varied by changing the angle of the incident beam while the laser wavelength is fixed. For example to make a 1D grating with period of 350 nm, the angle of incidence θ should be 27° given by Eq. 3.10. The two rotating mirrors were carefully aligned to give incidence angles of ±27°, and the incoming two beams were perfectly overlapped in the centre of the sample.

Typically the UV exposure took a few 10s of seconds, determined by the He-Cd laser power density. After exposure the sample was developed in a solvent MF319 for 1–2 s, to form micro-structures in the resist layer. The resulting structures were then characterized using an atomic force microscope (AFM), as shown in Fig. 3.17b. The AFM is described in more detail in Sect. 3.5.6. To transfer the pattern into glass for DFB laser fabrication, a reactive-ion etching (RIE) process was required, see Fig. 3.18, which is presented in detail in Sect. 3.5.5. The UV exposure and develop time were critical for optimization of the grating quality.

3.5.3 Electron Beam Lithography

Two-beam holography is used to fabricate large-area line-gratings in one step, whereas electron beam (e-beam) lithography is a more complicated technique that can define 'any-shape' gratings, although the fabrication time can be very long (10 s of hours for a 1 by 1 cm area). Thus e-beam lithography is usually limited to the production of relatively small patterns, however, it can produce much improved grating uniformity compared to the holographic technique. The resolution of an e-beam system is also significantly enhanced, by the simple fact that the wavelength

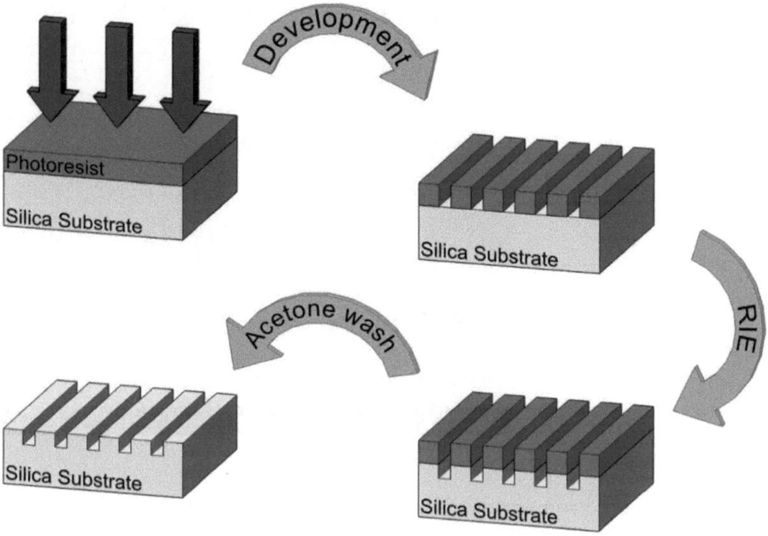

Fig. 3.18 Illustration of pattern transfer from resist into substrate

of electrons ($\lambda_{electron}$ ~ 12 pm, at acceleration voltages of 10 kV) is much smaller than of photons (λ_{photon} ~ 300 nm for UV illumination) used in optical lithography techniques. However, the actual resolution of the e-beam lithography is further limited by the forward scattering of the electrons as they pass through the film and by the backward scattering of these electrons from the substrate below the film. With the system at St Andrews, features as small as 20 nm in diameter are routinely fabricated.

The first step towards e-beam fabrication was the design of a pattern, which was accomplished using CAD (Computer-aided Design) software such as WaveMaker (WAM) or Elphy Plus (proprietary software supplied with the e-beam lithography system). The software generated a GDSII formatted file that the e-beam system could interpret. The electron beam was then focussed onto a resist, achieving a spot size less than 20 nm in diameter; this spot is then raster-scanned across the substrate exposing the areas of the software-generated pattern. The e-beam gratings used in this thesis were a mixture of in-house gratings and gratings fabricated at Kelvin Nanotechnology (Glasgow).

For in-house fabricated structures, the pattern was written into a thin positive e-beam resist layer (ZEP 520A), which was first spin-coated on to a substrate (typically 3,500 rpm for 60 s), usually of quartz or silicon, and baked at 180 °C for 10 min before e-beam exposure. The sample was then exposed using an appropriate basic step size (e.g. 2 nm) with a settling time (the time the system waits between moving the beam and exposing the particular portion of the pattern) of at least 1 ms. Typically an area dose of 55 μAcm^{-2} was used in the e-beam exposure. After exposure, the sample was developed in xylene (at 23 °C) for 45 s; this

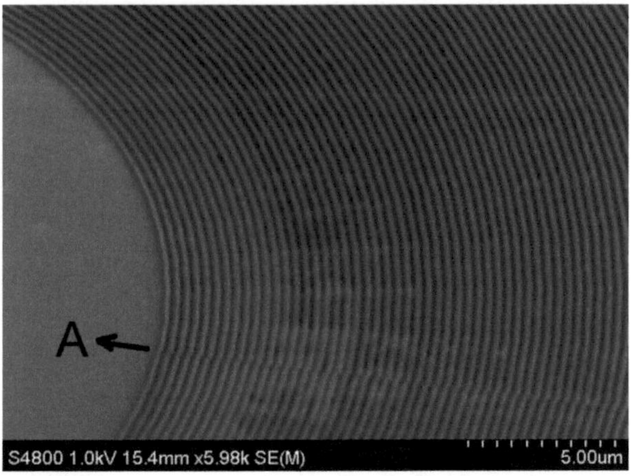

Fig. 3.19 Image of a ring grating obtained with a scanning electron microscope (SEM); *A* in the figure points out the stitching error

washed away the electron-exposed areas of the ZEP resist. This ZEP pattern was then used as an etch mask in a subsequent pattern transfer process, i.e. RIE. An example of an e-beam fabricated circular grating in silicon is shown in Fig. 3.19.

In order to expose a large-area grating, a number of write-areas must be used. The writing area of the e-beam system is defined as the largest area that can be exposed without moving the substrate, in other words, it is determined by the maximum deflection angle of the focussed e-beam. For the e-beam system in St Andrews, to overcome this limitation, multiple write-areas were "stitched" together to form a large exposure area. After each write-area finished, the stage was moved exactly one write-area horizontally or vertically. If a misalignment occurred between the stage and the electron beam position, defects at these write-area boundaries could become evident, these defects are known as stitching errors, see position A in Fig. 3.19. To minimize stitching error, precise alignment of the electron beam and the writing stage was essential.

3.5.4 UV Nanoimprint Lithography

A key advantage of nanoimprint lithography (NIL) is its simplicity, leading to low-cost fabrication of large-area nano-structures on a range of substrates including flexible plastic. It can be imagined as a imprinting tool combined with optical lithography, which can also have high resolution down to a few 10 s nm.

Figure 3.20 illustrates a schematic of UV-NIL. An EVG620 photomask aligner with custom tooling for UV-NIL (EV Group, Austria) was used in my PhD work. A high resolution master was fabricated by electron beam lithography, which was then used to cast a soft stamp on a transparent substrate such as glass. The master wafer was vapour-coated with a monolayer of anti-sticking material, per-fluoro-octyl-trichlorosilane (Sigma Aldrich), to make it highly hydrophobic.

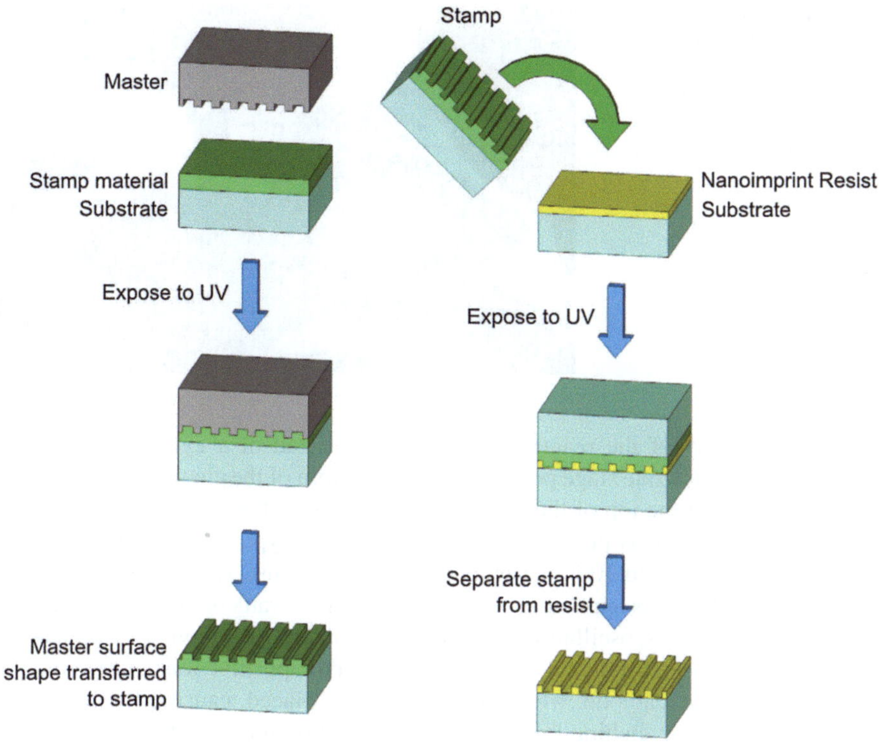

Fig. 3.20 Schematic diagram of UV nanoimprint lithography

The soft stamp was made from perfluoro polyether following a proprietary process of EV Group. With hard contact made between the stamp and the master with a high-dose UV exposure for 200 s, the pattern was transferred from the master to the UV-cured polymer stamp. As the master surface was highly hydrophobic, it could be separated from the stamp without polymer residue. The soft stamp, for example as shown in Fig. 3.21, was then used to print the nanostructures into a UV-NIL resist mr-UVCur06 (Microresist Technologies) by vacuum contact and UV curing, a soft stamp could be used for multiple imprinting processes. More details are described in Chap. 6.

3.5.5 Reactive Ion Etching

As mentioned above, reactive ion etching (RIE) was used to transfer the nanostructures from an etching mask, such as a photoresist layer, an e-beam resist layer or a nanoimprint resist layer, into a substrate, which could be quartz or silicon with high reproducibility.

Fig. 3.21 Photograph of a soft stamp with multiple grating areas, made with nanoimprint lithography (glass substrate was removed from the stamp)

The main body of the reactive ion etcher consisted of a cylindrical vacuum chamber with a substrate stage positioned at the bottom of the chamber. Gas entered through an inlet in the top of the chamber, and exited to the vacuum pump system through the bottom. Gas pressure in the chamber was typically maintained in the range between 10^{-6} and 10^{-4} bar by adjusting the exhaust. A radio-frequency (rf) field of 13.56 MHz was applied at between 20 and 100 watts between two parallel-plate electrodes. This oscillating electric field ionized the input gas molecules therefore creating a plasma. A DC bias accelerated the positive ions of the reactive plasma towards the cathode, where the sample was placed, causing ion bombardment and anisotropic etching (vertical etch rate $\gg$ horizontal etch rate).

Reactive gases, namely CHF_3, SF_6 and O_2 were used during the etching processes depending on the substrates to be etched. Etching rates depended on many processing parameters, such as chamber pressure, gas flow, rf power and the target material. Generally speaking for silicon etching, a gas mixture of 1:1 CHF_3 to SF_6 with an rf power of 20 W is used to achieve an approximately 100 nm/min etch rate, while for a silica etch, CHF_3 with an rf power of 40 W could produce an etching rate of approximately 10 nm/min. The depth of the etched gratings was controlled by the etching time. Figure 3.22 shows a scanning electron microscopy (SEM) image of an etched grating in a silicon substrate, indicating the depth of the grating to be approximately 75 nm.

3.5.6 Microstructure Characterization

Scanning electron microscopy was one of the techniques used to examine the morphology of the micro-structures, however, in order to measure the groove depth, the sample needed to be cleaved to expose the facet of the grating. In contrast, atomic force microscope (AFM) was another high resolution scanning microscope which could inspect the grating surface and groove depth non-destructively.

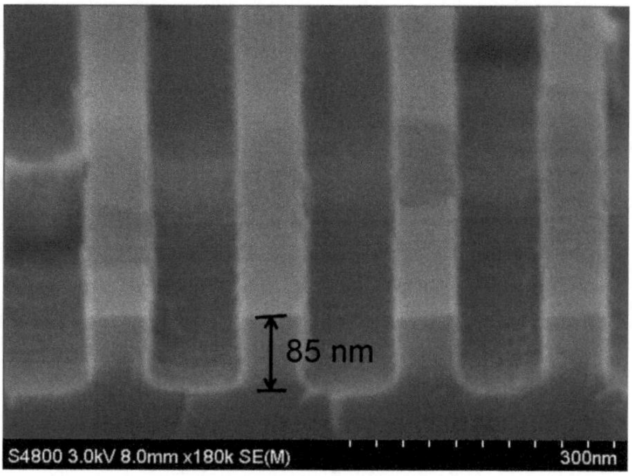

Fig. 3.22 An SEM image of the facet of a cleaved grating tilted at 25°, indicating roughly 75 nm grating depth

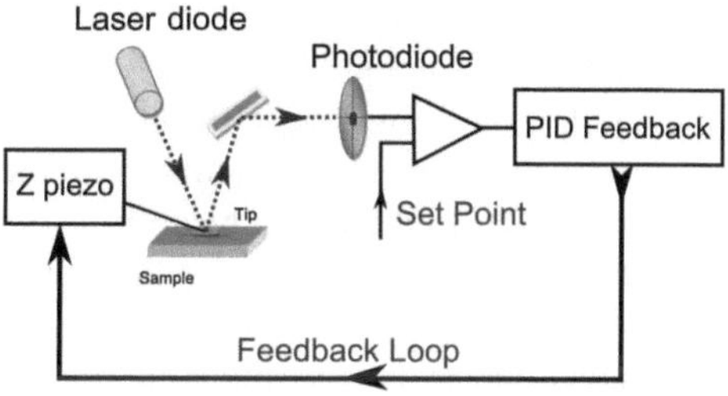

Fig. 3.23 Signal pathway for contact mode of AFM

Furthermore, the AFM is also a very useful technique to measure the surface roughness of polymer films, phase separation in a blend, the thin-film thickness and so on.

A Caliber AFM (Veeco Instruments Inc.) was used in this work. It could be operated in two modes, i.e. contact and tapping mode. Contact mode is for generic samples and tapping mode is designed for very soft surfaces and phase measurements. Here I take the contact mode as an example to explain the operation of the AFM.

The AFM consisted of a micro-scale cantilever with a sharp tip at its end which was used to scan the sample surface. When the tip was brought into proximity of a surface, the force between the tip and the sample molecules led to a deflection of the cantilever. A CW laser diode was placed in the AFM head, along with a mirror and a photodetector, as shown in Fig. 3.23. By adjusting these three elements, the reflected laser beam from the cantilever tip was directed towards the centre of a detector, with an electronic signal fed back into the circuit. The feedback loop aimed to keep a constant force/distance between the cantilever and the sample surface. In this way, the scanning image reflected the topology of the sample surface.

3.6 Organic Semiconductor Laser Characterization

3.6.1 Laser Threshold

An organic semiconductor DFB laser can be fabricated by spin coating the gain medium onto the corrugated substrate. In this thesis, the DFB lasers all emit at an angle normal to the surface of the laser.

Several properties of the organic semiconductor laser (OSL) were routinely characterized, including lasing threshold and wavelength. The experimental setup is shown in Fig. 3.24, which is similar to the ASE threshold measurement. To avoid degradation, the laser was either placed inside a vacuum chamber or encapsulated with a capping layer such as CYTOP (Asahi Glass, Japan), an optically-transparent polymer. The pump laser could be replaced with other lasers including micro-chip lasers and laser diodes. The pump intensity was varied with neutral density filters. Using inorganic LEDs to pump the OSL is discussed separately in Chaps. 5 and 6.

A CCD fibre-coupled detector was placed at one side of the chamber to collect the DFB laser emission at the normal angle, while a sensitive photodiode was placed on the opposite side to measure the OSL output energy. The OSL was tilted at roughly 20° to the pump beam to avoid the reflected pump beam being directly incident onto the detector. As the pump energy increased, a sharp peak in the emission spectrum was observed, accompanied by a sudden change in the slope of the output energy against input energy, indicating the start of lasing, see Fig. 3.25. A slope efficiency above threshold, ρ can be calculated by:

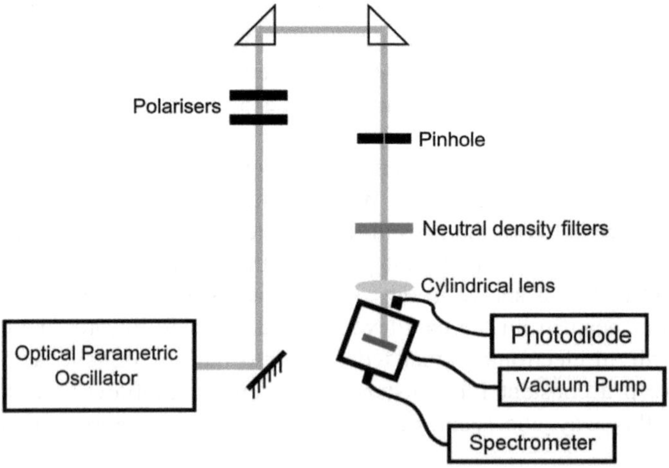

Fig. 3.24 Experimental set-up for organic laser characterization

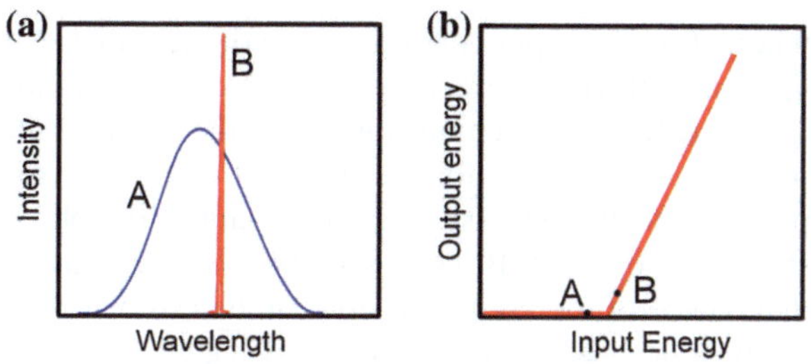

Fig. 3.25 Illustration of **(a)** spectrum below (A, *blue line*) and above threshold (B, *red line*); **b** the output power as a function of the input pump power, with a 'kink' at the threshold [14]

$$\rho = \frac{OuputEnergy}{InputEnergy - ThresholdEnergy} \times 100\,\% \tag{3.11}$$

3.6.2 Angle Dependent Emission

The strongly *angle-dependent emission* for a surface-emitting DFB laser is a useful property that can provide insight to the properties of the guided modes and the divergence of the light emission. Figure 3.26 presents a simplified set-up of the angle-dependent measurement. Emission spectra were collected with a fibre or a detector (aperture < 1 mm) at a known distance, in the range of $\pm 5°$ ($0°$ is defined at the perpendicular direction to the OSL).

As discussed in Chap. 2, different wavelengths of the spontaneous emission couple into different waveguide modes propagating in the waveguide with wavevector k_{mode}. The phase matching condition in the plane of the film can be written as:

$$k_{//} = k_{\text{mode}} + mG \tag{3.12}$$

where m is a diffraction order and G is the grating vector. This can be rewritten as:

$$\frac{2\pi}{\lambda}\sin\theta = \pm\frac{2\pi}{\lambda}n_{eff} + m\frac{2\pi}{\Lambda} \tag{3.13}$$

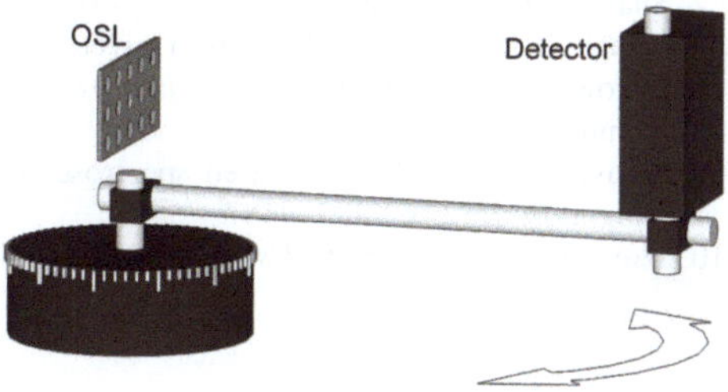

Fig. 3.26 Experimental set-up for *angular-dependent emission* characterization

where θ is the output coupling angle of the radiative mode with a wavelength at λ, n_{eff} is the effective refractive index of the guided mode and Λ is the gating period. This equation also demonstrates how the angle of the scattered light relates to the in-plane wave-vector for the different emission wavelengths. By plotting a mapping of detected intensity against λ and θ, the photonic mode dispersion and the stop-band of a DFB laser can be observed, as described in Sect. 2.5.1. The divergence of the laser emission can be estimated with a known distance between the polymer laser and the detector. Chapter 5 shows more experimental results on the angle-dependent measurements.

3.7 Conclusions

This chapter has focused on the photophysical and gain characterization of organic semiconductors, and the fabrication techniques used for making optically pumped organic lasers. The rest of this thesis consists of five chapters on experimental results and one on conclusions and future work. The theories and experimental methods on explosive sensing are discussed separately in Chaps. 7 and 8.

References

1. Hall, D. B., Underhill, P., & Torkelson, J. M. (1998). Spin coating of thin and ultrathin polymer films. *Polymer Engineering & Science, 38*(12), 2039–2045.
2. Madou, M. J. (2002) *Fundamentals of microfabrication and nanotechnology* (3rd ed.). Boca Raton: CRC Press.
3. Scriven, L. E. (1988). Physics and applications of dip coating and spin coating. *MRS Proceedings, 121*, 717–729.
4. Varian Inc. Cary® 100/300 UV-Vis Spectrophotometers. http://www.bu.edu/chemistry/files/cic/other/Varian%20UV%20Intro.pdf
5. Greenham, N. C., Samuel, I. D. W., Hayes, G. R., Phillips, R. T., Kessener, Y. A. R. R., Moratti, S. C., et al. (1995). Measurement of absolute photoluminescence quantum efficiencies in conjugated polymers. *Chemical Physics Letters, 241*(1–2), 89–96.
6. Suzuki, K., Kobayashi, A., Kaneko, S., Takehira, K., Yoshihara, T., Ishida, H., et al. (2009). Reevaluation of absolute luminescence quantum yields of standard solutions using a spectrometer with an integrating sphere and a back-thinned CCD detector. *Physical Chemistry Chemical Physics, 11*(42), 9850–9860.
7. Shaw, P. E. (2009). *Measurements of exciton diffusion in conjugated polymers*. Thesis for the degree of Doctor of Philosophy, Univeristy of St Andrews, St Andrews.
8. Lakowicz, J. R. (1991). *Topics in fluorescence spectroscopy volume 1 techniques*. New York: Plenum Press.
9. Meuse, C. W. (2000). Infrared spectroscopic ellipsometry of self-assembled mono layers. *Langmuir, 16*(24), 9483–9487.
10. *User Manual—Guide to Using WVASE32*. J. A. Woollam Co., Inc.

11. Maiti, P., Nam, P. H., Okamoto, M., Hasegawa, N., & Usuki, A. (2002). Influence of crystallization on intercalation, morphology, and mechanical properties of polypropylene/clay nanocomposites. *Macromolecules, 35*(6), 2042–2049.
12. McGehee, M. D., & Heeger, A. J. (2000). Semiconducting (conjugated) polymers as materials for solid-state lasers. *Advanced Materials, 12*(22), 1655–1668.
13. Lawrence, J. R., Andrew, P., Barnes, W. L., Buck, M., Turnbull, G. A., & Samuel, I. D. W. (2002). Optical properties of a light-emitting polymer directly patterned by soft lithography. *Applied Physics Letters, 81*(11), 1955–1957.
14. Samuel, I. D. W., Namdas, E. B., & Turnbull, G. A. (2009). How to recognize lasing. *Nature Photonics, 3*(10), 546–549.

Chapter 4
Low-Threshold and Broadly Tuneable Organic Lasers Based on Star-Shaped Oligofluorene Truxenes

4.1 Introduction

Star-shaped oligomers, defined as branched molecules that consist of linear oligomer arms joined together by a central core, have been of great interest in the past decade as possible alternatives to linear conjugated oligomers in various optoelectronic applications, including light-emitting devices [1], solar cells [2], and transistors [3]. A star-shaped architecture combined with their conjugated character within the arms brings new electrical, optical, and morphological properties to the system [4].

In this chapter a number of star-shaped gain materials which hold promise for use in organic lasers are studied. In Sect. 4.2 the individual materials that have been studied are listed and described. In Sect. 4.3 the photophysical properties including absorption, photoluminescence, PLQY and excited-state decay lifetime that have been studied for each material are presented, and in Sect. 4.4 the gain properties of these materials which were investigated using amplified spontaneous emission (ASE) are discussed. Results on variable-angle spectroscopic ellipsometry (VASE) measurements are presented in Sect. 4.5 including the optical constants of the gain media. In Sect. 4.6 surface-emitting distributed feedback (DFB) lasers are examined with different resonators, as well as the emission spectral tunability of these lasers. Towards the end of this chapter, the gain properties of macromolecules with different arm-configurations are discussed in Sect. 4.7, followed by the conclusions in Sect. 4.8.

4.2 Gain Media: Truxene T3 and T4

In contrast to polymers, oligomers are monodisperse and characterized by a well-defined molecular structure, and can exhibit superior chemical purity [4]. However, it has often been observed that one-dimensional (linear chain) conjugated oligomers can be prone to self-aggregation in the solid state (through π–π stacking) which can

Y. Wang, *Low Threshold Organic Semiconductor Lasers*, Springer Theses,
DOI: 10.1007/978-3-319-01267-4_4,

result in quenching of emission [5, 6]. This issue can be addressed by bonding several linear oligomers together, for example, attaching the linear oligomers to a truxene (10,15-dihydro-5H-diindeno[1,2-a;1′,2′-c]fluorene) core, which is composed of a polycyclic aromatic system with C3 symmetry [7].

The truxene core, shown in Fig. 4.1a, is crowded by six hexyl groups that are perpendicular to the fused aromatic rings; therefore aggregation is prevented by this multi-dimensional structure. Onto this core, arms consisting of oligofluorene units are attached with varying lengths leading to a family of truxene-fluorene oligomers, starting with the bare core (T0) and with arms consisting of one (T1), two (T2), three (T3) and four (T4) 9,9-dihexylflurorene units. Figure 4.1b shows the structures of the star-shaped oligofluorene family T1–T4. In the case of T4, the radius of the macromolecule is approximately 4 nm, which represents one of the largest star-shaped conjugated oligomers [4].

The truxene materials used in this work were synthesised by Dr. Alexander Kanibolotsky, Prof. Peter Skabara at the University of Strathclyde in Glasgow and

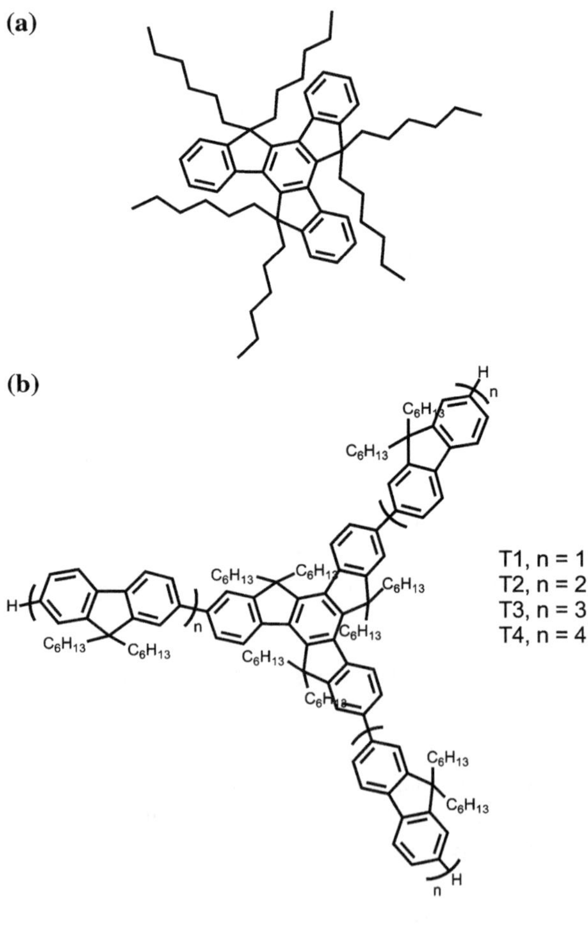

Fig. 4.1 Chemical structures of **a** a truxene core and **b** the truxene-oligofluorene family

Prof. Igor F. Perepichka at the University of Central Lancashire in Preston. In this investigation, the two longest-arm members of the family, T3 and T4, are of particular interest. Their unique molecular shape leads to several advantages including good solubility in common organic solvents and improved morphology of the thin films. The solution processing of these materials promises low-cost opto-electronic devices and simplified fabrication of wavelength-scale structures. This is particularly relevant for organic semiconductor lasers, as discussed in Chap. 2.

4.3 Photophysical Properties

The absorption and fluorescence spectra, plotted in Fig. 4.2, were measured from thin films, spin-coated from toluene solutions (20 mg/ml) on quartz substrates, using the Cary 300 UV–Vis and the FluoroMax-2 spectrometers respectively, following the methods described in Chap. 3. In Table 4.1 their absorption and photoluminescence (PL) emission (0–0, 0–1 and 0–2 transitions) peak-wavelengths are presented, as well as their solid-state photoluminescence quantum yield (PLQY), measured using the method of integrating sphere with a CW excitation at 325 nm from a He-Cd laser [8].

The spectra in Fig. 4.2 show that there is an approximate 6 nm red-shift in both absorption and emission of T4, which is due to the extra 9,9-dihexylflurorene unit in each arm compared to the T3 molecule. This increased conjugation length results in the decreasing of the gap between the highest occupied molecular orbital (HOMO) and the lowest unoccupied molecular orbital (LUMO), hence lowering the energy of the electronic transition, as explained in Chap. 2. From Table 4.1,

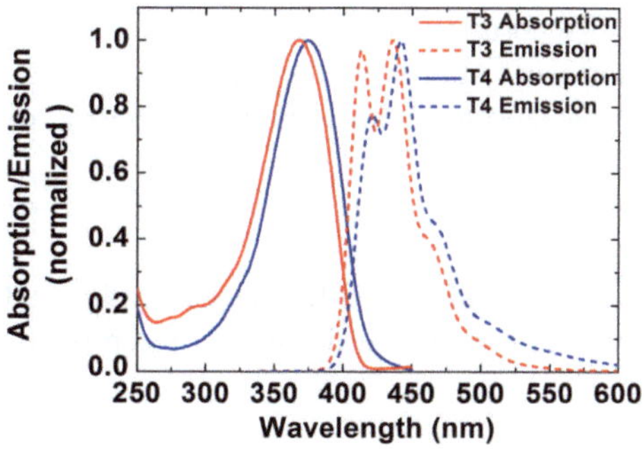

Fig. 4.2 Absorption (*solid line*) and photoluminescence (*dashed line* $\lambda_{exc} = 325$ nm) spectra of T3 (*red*) and T4 (*blue*) truxenes

Table 4.1 Photophysical properties of T3 and T4 in thin-films

	$\lambda_{abs\text{-}max}$ (nm)	$\lambda_{PL\ 0\text{–}0}$ (nm)	$\lambda_{PL\ 0\text{–}1}$ (nm)	$\lambda_{PL\ 0\text{–}2}$ (nm)	PLQY[a] (%)
T3	367	414	437	461	86
T4	375	420	441	467	73

[a] Estimated uncertainty in PLQY is ±10 %

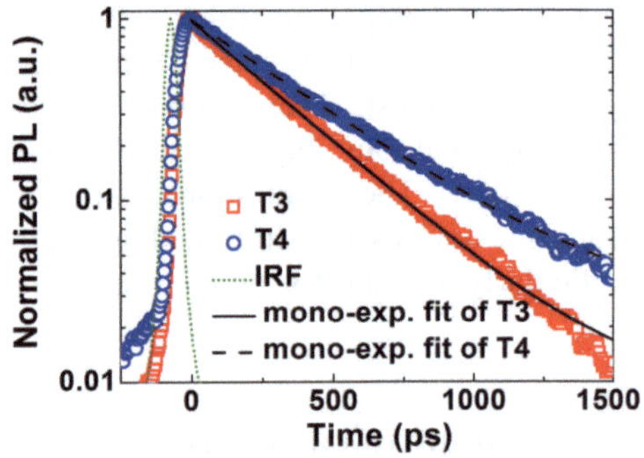

Fig. 4.3 Dynamics of PL decays in T3 (*red open squares*) and T4 (*blue open circles*) films are recorded; the *green dotted line* indicates the instrument response function (IRF, FWHM of 40 ps); the *solid line* indicates the mono-exponential fit for T3, and the *dashed line* indicates the mono-exponential fit for T4

we see that both of these two longest-arm members in the truxene-oligomer family have very high solid-state PLQY, compared to the PLQY value of the linear fluorene polymer poly(9,9-dioctylfluorene) (PFO) which has been reported to be typically 50 % [9] and linear oligofluorenes (O3 and O4) reported to be about 65 % [10]. In Montgomery et al's work [10], direct comparisons of the photophysics confirmed that the star-shaped molecules have larger molar absorption coefficients and higher PLQYs than straight chain oligofluorenes, which make them more attractive for light-emitting devices such as lasers.

The photoluminescence lifetime of T3 films was measured with a streak camera in a spectral window of 400–500 nm. By fitting the time-resolved PL decay from the T3 film with a mono-exponential decay curve:

$$I(t) = I_0 e^{-t/\tau} \tag{4.1}$$

A singlet excited-state lifetime of 330 ± 20 ps was measured. A slightly longer excited-state lifetime of 400 ± 20 ps was measured for T4 films, as shown in Fig. 4.3.

Accordingly using Eq. 4.2, the radiative lifetimes of T3 and T4 can be deduced, and were approximately 380 and 540 ps, respectively. The glassy phase PFO has been reported with a radiative lifetime of 540 ps [11].

$$\tau_{rad} = \frac{\tau_{PL}}{PLQY} \tag{4.2}$$

Short radiative lifetime generally lead to a reduction in the losses associated with non-radiative decay and the probabilities of triplet–triplet annihilation.

4.4 ASE Studies of T3 and T4

To further investigate the potential of T3 and T4 truxenes as laser gain media, amplified spontaneous emissions (ASE) were studied in thin-film waveguides. A good laser gain material should offer high optical gain per unit length, as this allows stronger amplification of the emitted light and potentially reduces the lasing threshold. In addition, the material should also have low optical loss as this would significantly affect the lasing threshold and efficiency.

4.4.1 ASE Thresholds

Thin films of T3 Truxene solution in toluene (20 mg/ml) were deposited on quartz slides to form air-T3-silica waveguides. The thicknesses of the films were 145 ± 20 nm, measured using the Veeco Dektak 150 surface profiler, which was sufficient to support the fundamental transverse electric mode (TE_0). The optical excitation pulses were generated by a Nd:YAG laser-pumped OPO with 4 ns output pulses at a 20 Hz repetition rate. The excitation wavelength was at 375 nm which was close to the absorption peak of T3. Calibrated neutral density filters were used to vary the pump intensity incident on the films. The emission spectra were collected from the edge of the film at various pump intensities with a constant pump stripe of dimension 4.5 by 0.2 mm, as shown in Fig. 4.4.

As the pump intensity was increased, the edge-emission spectrum of T3 became narrower, changing from a broad PL to a waveguided ASE spectrum with a maximum intensity at 439 nm, near the 0–1 vibronic peak of its PL (see Fig. 4.4), where the net gain was maximal in the film. Figure 4.5 shows the relative output intensity, integrated over the ASE peak, as a function of the pump intensity, along with the full width at half-maximum (FWHM) linewidth of the edge-emission spectrum at different pump intensities. The FWHM linewidth dropped dramatically to 4.6 nm when the pump intensity was increased above 10.6 μJ/cm^2, equivalent to 2.1 kW/cm^2 [12]. At the same time, the output emission intensity increased rapidly. Thus an ASE threshold as low as 2.1 kW/cm^2 was measured, which was substantially lower than the corresponding value of 58 kW/cm^2 reported for PFO [13].

The ASE threshold in an air-T4-silica waveguide was also measured. With a 130 nm thick film, the ASE peak was observed at the wavelength of 440 nm. The ASE threshold was found to be 4 kW/cm^2, as shown in Fig. 4.6, with a pump

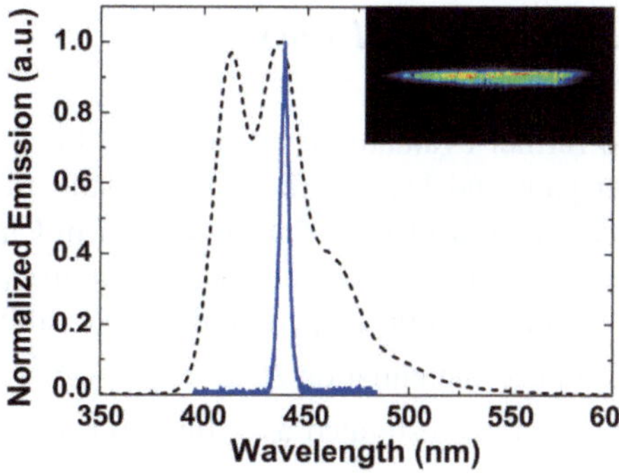

Fig. 4.4 Photoluminescence (*dashed line*) and amplified spontaneous emission (*solid line*) spectra of T3 truxene; *Inset* intensity profile of the excitation beam in the ASE measurement

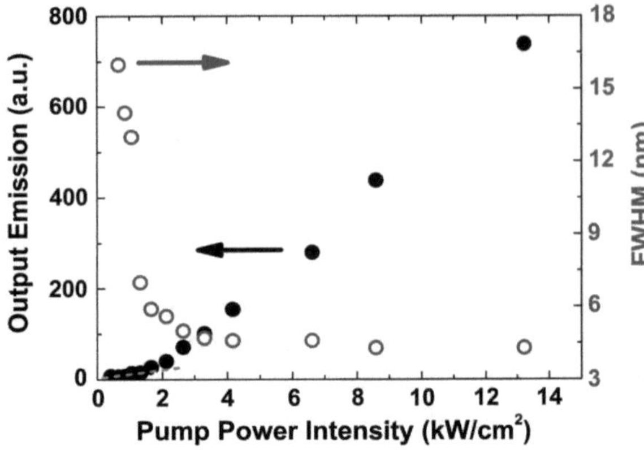

Fig. 4.5 The dependence of the total emission intensity from the edge of the T3 waveguide (*closed circles*) and the FWHM of the spectra (*open circles*) on the pump power intensity; the *dashed line* is a linear fitting for indication of the pump threshold

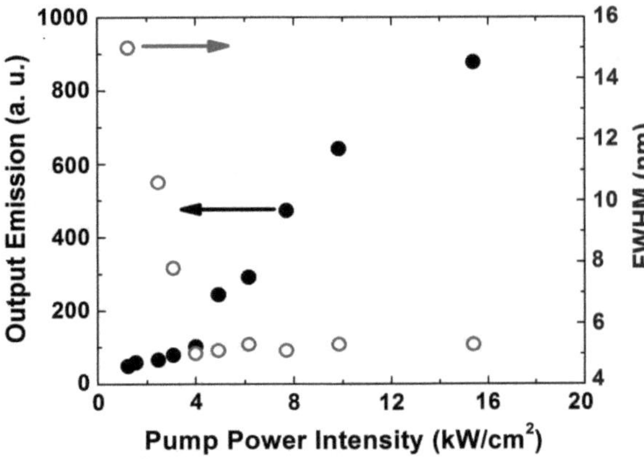

Fig. 4.6 The dependence of the edge-emission intensity from the T4 waveguide (*closed circles*) and the FWHM (*open circles*) on the pump power intensity

stripe of 2.3 by 0.1 mm from the same excitation source as in the T3 ASE measurements. These low ASE thresholds of both T3 and T4 indicate there effective amplifications of light in these truxene thin-films.

4.4.2 Gain and Loss

To further evaluate the potential of this material as a laser gain media, waveguide net gain and loss coefficients are studied for both T3 and T4 using the variable stripe length method that described in Chap. 3.

Figure 4.7a shows the output intensity at the peak of the ASE spectrum as a function of pump stripe length at five different pump energy levels. By fitting each dataset to the equation $I(\lambda) = \frac{A(\lambda)I_p}{g(\lambda)}(e^{g(\lambda)l} - 1)$, the net gain of the T3 waveguide, *g(λ)*, was investigated as a function of pump intensity. Figure 4.7b indicates that

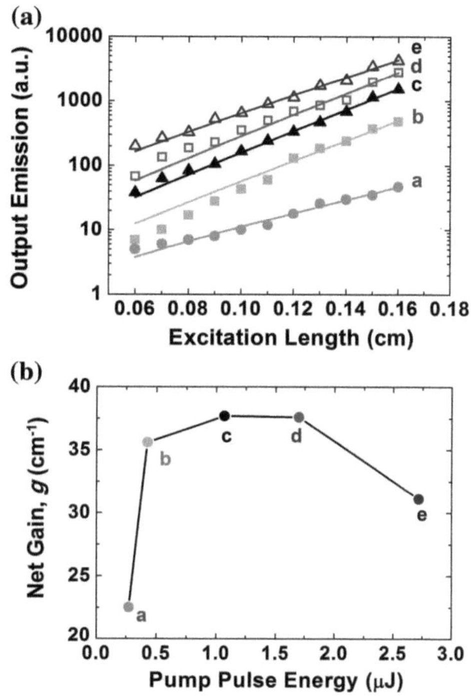

Fig. 4.7 a Emission intensity (above ASE threshold) of a T3 waveguide as a function of excitation length at five different pump intensities; *a* 0.27 μJ/pulse, *b* 0.43 μJ/pulse, *c* 1.1 μJ/pulse, *d* 1.7 μJ/pulse, *e* 2.7 μJ/pulse; **b** the relationship between gain coefficient and excitation energy; the gain coefficients from the fitted data are: *a* 23 cm^{-1}, *b* 36 cm^{-1}, *c* 38 cm^{-1}, *d* 38 cm^{-1}, *e* 31 cm^{-1}

g increases with the pump energy at low levels. At higher pump energy, the gain coefficient stopped increasing with the pump energy, thereafter, it was even seen to decrease. This indicated the presence of gain saturation in the truxene waveguide. It occurs when the light travelling in the waveguide is too intense so as to deplete a substantial fraction of the excitons; also at the high pump intensity, the exciton–exciton annihilation would rapidly decrease the exciton density resulting in a reduction of the gain coefficient [14, 15]. The highest waveguide gain coefficient of T3 found in the experiments was 38 cm^{-1} with pump energy of 1.1 μJ/pulse.

To measure the loss, instead of changing the length of the pump stripe, the pump stripe was moved away from the edge of the film. Figure 4.8 shows the ASE peak intensity as a function of *x,* the distance between the end of the pump stripe and the edge of the film. By fitting the data to an exponential decay, $I_{\text{output}} = I_{\text{pump}}e^{-\alpha x}$, an average loss coefficient, α, was measured to be approximately 3.3 cm^{-1}. Table 4.2 summarizes the gain and loss coefficients measured with T4 truxene films. The maximum gain for T4 was 23 cm^{-1}, and the average loss coefficient was 2.3 cm^{-1}.

Compared to the lowest optical loss values reported for solution-processed neat films of fluorene-based organic semiconductors which are in the range of 3.5–6 cm^{-1} [16–18], the loss coefficients for T3 and T4 films are very low. This result indicates that the light travelling within the thin-film waveguides experiences very little loss, either from self-absorption or scattering.

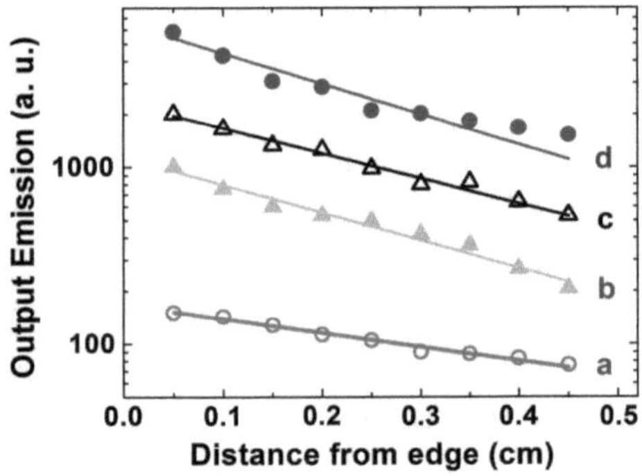

Fig. 4.8 The ASE intensity of a T3 film at four different pump energies as a function of the distance between the pump stripe and the edge of the film; the loss coefficients from the fitted data are: *a* 2.3 cm^{-1}, *b* 3.6 cm^{-1}, *c* 3.3 cm^{-1}, *d* 4.0 cm^{-1}

Table 4.2 Gain and loss in an air-T4-silica waveguide

	Excitation energy(μJ/pulse)	Fitted value (cm^{-1})
Gain	0.17	13
	0.27	20
	0.43	23
	0.68	23
Loss	0.19	2.2
	0.30	2.3
	0.48	2.3
	0.60	2.2

Considering the above improvements in key photophysical properties including PLQY, gains and losses, T3 and T4 truxenes have shown great potential as organic laser gain media.

4.5 Optical Constants

Before making a laser, the optical constants of T3 and T4 thin-films were studied using variable-angle spectroscopic ellipsometry (VASE) for angles between 45° and 75° in steps of 5°. A set of T3 films with different thicknesses (30–40 nm) were prepared by spin coating from a T3 solution (toluene, 10 mg/ml) on silica substrates. Ellipsometry measurements were performed using a J. A. Woollam M2000-DI ellipsometer, as discussed in Chap. 3. Transmission measurements with incidence light at 90° were also performed and the transmission data were attached to the reflection data for each sample respectively. The data were first fitted with an isotropic Cauchy model between a spectral window of 450–1700 nm (T3 films are transparent in this wavelength range) to determine the film thicknesses, using the *WVASE32* software. Knowing the film-thicknesses, the optical constants, i.e. the refractive index n and the extinction coefficient κ, were obtained by a point-by-point fit to the data in the whole spectral range of 190–1700 nm. Ellipsometry measurements were performed with a set of T4 thin-films as well.

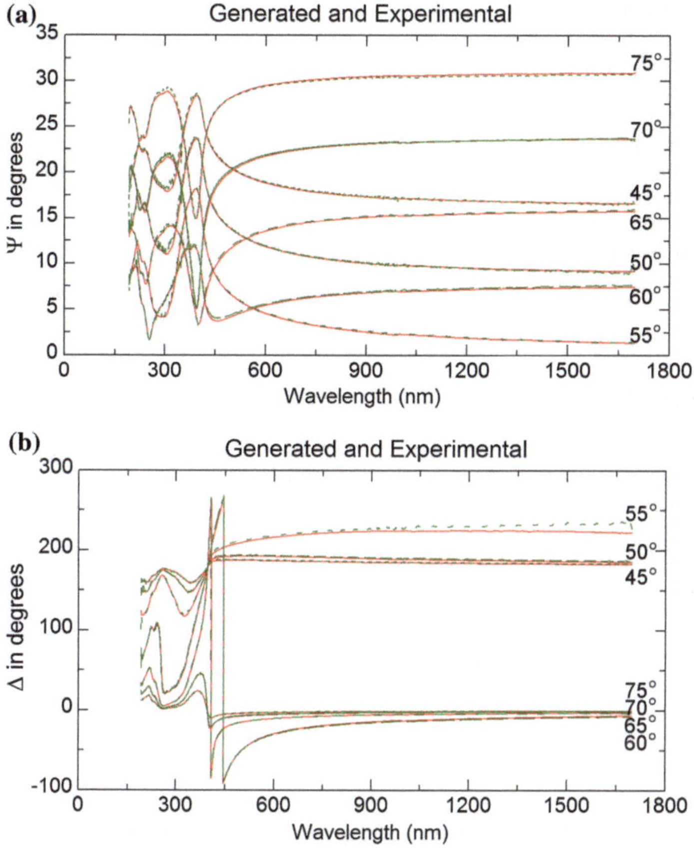

Fig. 4.9 Ellipsometry experimental data **a** Ψ and **b** Δ and fitted data to an isotropic model for angle of incidence from 45° to 75°, obtained from a 40 nm thick T3 truxene film

Here I take the T3 as an example. Figure 4.9 shows the experimental ellipsometry data, Psi (Ψ) and Delta (Δ), and the isotropic model generated data. These two show an excellent agreement with a mean squared error (MSE) of 0.8879. Uniaxial fittings were performed to confirm that the truxene films were isotropic. Figure 4.10 represents high agreement between the refractive indices in ordinary and extraordinary directions. This confirmed that the T3 films are isotropic, i.e. the molecules arrange themselves in random directions within the film. The random arrangement of the molecules was also a sign of the absence of crystalline phases. This result was consistent with the small waveguide losses observed for these truxene-oligomer films in the ASE measurements.

Furthermore, the optical constants of both T3 and T4 films are plotted between 200 and 700 nm in Fig. 4.11. The optical constants for T3 and T4 are very similar, with a few-nm red-shift for T4 as expected. T4 has slightly higher maximum extinction coefficient due to its longer arms, i.e. higher absorption coefficient.

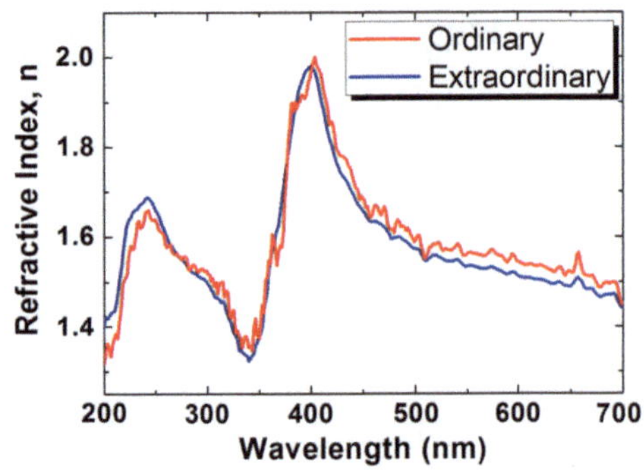

Fig. 4.10 Refractive indices of T3 films in ordinary (*red*) and extraordinary (*blue*) directions obtained from an anisotropic fitting; the similarity in the refractive indices in these two directions indicates the T3 films are isotropic; and the rough curves also imply poor fitting with the anisotropic model

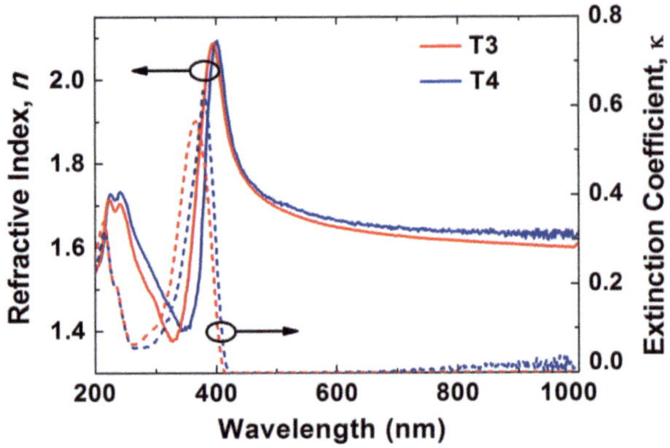

Fig. 4.11 Optical constants of T3 (*red*) and T4 (*blue*); n_{T3} at 440 nm is 1.78; n_{T4} at 440 nm is 1.80; κ_{T3} is peaked at 367 nm with a value of 0.57 while κ_{T4} is peaked at 378 nm of 0.63

The real parts of the refractive indices at around 440 nm were used in the calculation of the appropriate film thicknesses for designing the distributed feedback lasers.

4.6 Laser Threshold and Tunability Measurement

Knowing the refractive index of the material, the effective refractive index n_{eff} for a guided transverse mode TE_0 at certain wavelength can be calculated using the equation below,

$$\frac{2\pi d}{\lambda}\sqrt{n_{T3}^2 - n_{eff}^2} = \tan^{-1}\sqrt{\frac{n_{eff}^2 - n_{air}^2}{n_{T3}^2 - n_{eff}^2}} + \tan^{-1}\sqrt{\frac{n_{eff}^2 - n_{sub}^2}{n_{T3}^2 - n_{eff}^2}} \tag{4.3}$$

where d is the film thickness, n_T and n_{sub} are the refractive indices of T3 and the substrate respectively. It is an important parameter for designing DFB lasers, as explained in Chap. 2. The effective refractive index of the fundamental TE mode was calculated for the air-T3-silica waveguide with the film thickness of 150 nm.

Taking the refractive index of silica, $n_{silica} = 1.46$ and $n_{air} = 1$, n_{eff} was then calculated to be 1.59 using Eq. 4.3. The Bragg wavelength was determined by the Bragg condition, $m\lambda_{Bragg} = 2n_{eff}\Lambda$, with a known period of the grating, for example 270 nm, for a second order surface emitting laser, i.e. m = 2. Hence the laser should emit at around 429 nm, perpendicular to the plane of the gratings.

The DFB laser was made with a 30 mg/ml T3 truxene solution, spin-coated onto a 1D silica grating, which was fabricated using the two-beam interference holography, with periods of 270 nm and groove depth of roughly 25 nm. An AFM image of the grating is shown in Fig. 4.12a. The sample was placed within a vacuum chamber evacuated to 10^{-4} bar and optically pumped with 355 nm pulses from a frequency-tripled Nd:YVO_4 laser (10 ns pulse-width at 500 Hz repetition rate). The pump beam was focused to a 0.11 cm spot in diameter on the sample, with the pump beam incident onto the sample at 20°. The laser emission was detected at the normal angle to the sample using the fibre coupled CCD spectrograph. Figure 4.12b shows a minimum laser threshold of 515 W/cm^2 with the lasing wavelength of 430 nm, and this wavelength agrees very well with the

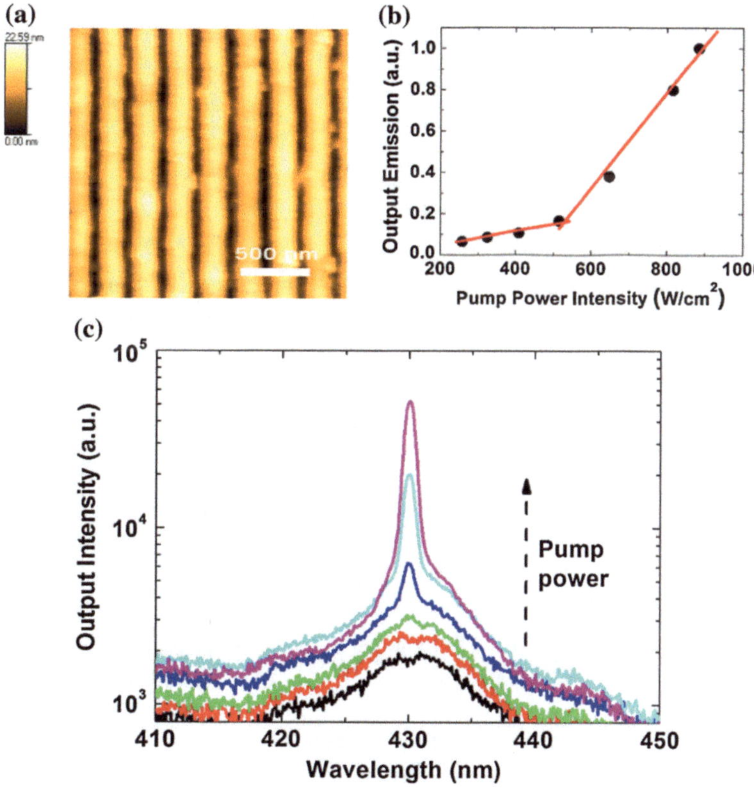

Fig. 4.12 **a** AFM image of the 270 nm period 1D grating; **b** T3 laser threshold of 515 W/cm^2 at emission wavelength of 430 nm; **c** emission spectra below and above laser threshold

calculated value above. The sudden change in the slope of the output as a function of pump intensity indicate the start of laser. Figure 4.12c illustrates the surface emission spectra below and above threshold.

To optimize the T4 lasing wavelength, it required a grating with slightly longer periods. Thin films were deposited from T4 toluene solutions on top of a holographic 2D silica grating with period of 300 nm and groove depth of 35 nm, see Fig. 4.13a, and excited with a focused beam (0.16 cm in diameter) from the same pump source used in the T3 laser measurements. Lasing from the T4 truxene laser was observed at a very low threshold of 270 W/cm^2 at the peak wavelength of 446 nm; [19] it is more than an order of magnitude lower compared to reported PFO 2D DFB laser with threshold of 3.2 kW/cm^2 [20].

The quality factors of these 1D and 2D DFB gratings are estimated to be above 100, based on the measured linewidth of the laser emission spectra. The quality factor is defined as:

$$Q = \frac{f_0}{\delta f} \tag{4.4}$$

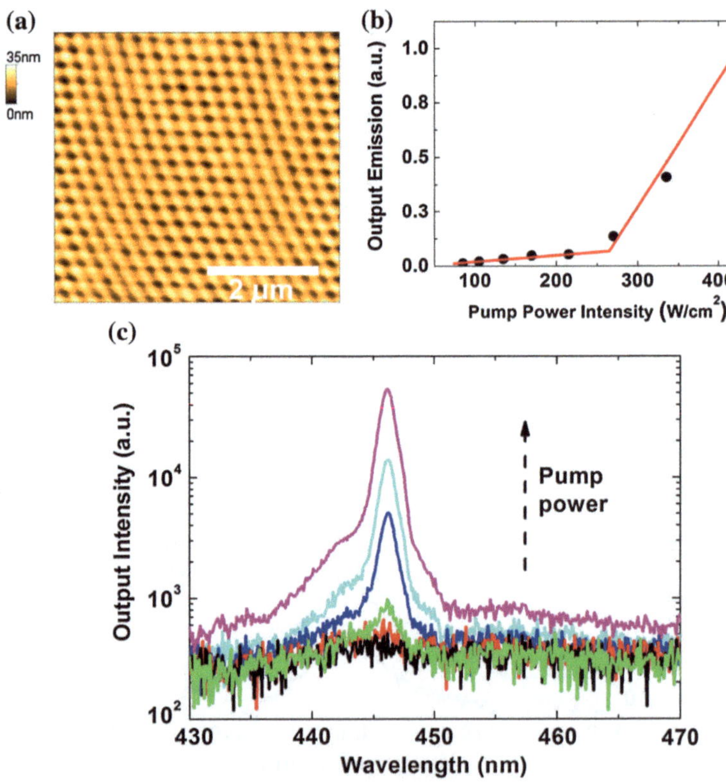

Fig. 4.13 **a** AFM image of the 300 nm 2D grating; **b** T4 laser threshold of 270 W/cm^2 at 430 nm emission wavelength; **c** emission spectra below and above laser threshold

where f_0 is the resonate frequency and δf is the FWHM of the bandwidth of the DFB resonator.

As mentioned in Chap. 1, one of the most attractive properties of organic semiconductor gain materials is that they have broad emission spectra. With a single gain medium, the organic DFB laser can be tuned by changing the film thickness or using a different grating period.

To make tunable T3 lasers, several silica gratings with periods of 270 nm (1D), 280 nm (1D) and 300 nm (2D) were used. Also, by changing the spin-coating speed, the thicknesses of T3 films were varied from 100 to 160 nm. For some T3 lasers, with films thicker than 100 nm, two lasing modes above threshold at the same time were observed. These were separated clearly in wavelength and were identified as lasing in the TE_0 and TM_0 modes, with a blue shift of TM mode against TE mode due to its lower effective refractive index. In addition, the laser thresholds were not the same for different wavelengths. The lowest threshold appeared with a lasing wavelength close to the ASE peak. Towards shorter wavelengths the optical loss increases due to the self-absorption, hence the lasing threshold rises. For longer wavelengths, the threshold increases due to the decrease of the wavelength dependent emission cross-section.

The T3 laser wavelengths span the range from 422 to 473 nm (2.62 to 2.94 eV), as shown in Fig. 4.14. This tuning range of 51 nm (0.32 eV) compares very favourably with other blue-emitting organic semiconductors, such as the tuning range of 39 nm (0.24 eV) reported for polyfluorene [20] and 33 nm (0.23 eV) for spirobifluorene [21]. Riechel et al. have reported a surface-emitting laser made with a thin film of tris(8-hydroxyquinoline) aluminium (Alq_3) doped with a laser dye 4-(Dicyanomethylene)-2-methyl-6-(julolidin-4-yl-vinyl)-4H-pyran (DCM2) has over 100 nm tuning range in the red spectral region [22]. Such wide tunability in a single material is an extremely attractive feature for numerous laser applications such as spectroscopy, lab-on-a-chip devices and data communication [23–25].

By changing the period of the corrugated substrate and also varying the film thickness, the tunability of the T4 laser wavelength was over 25 nm, from 428 to 453 nm.

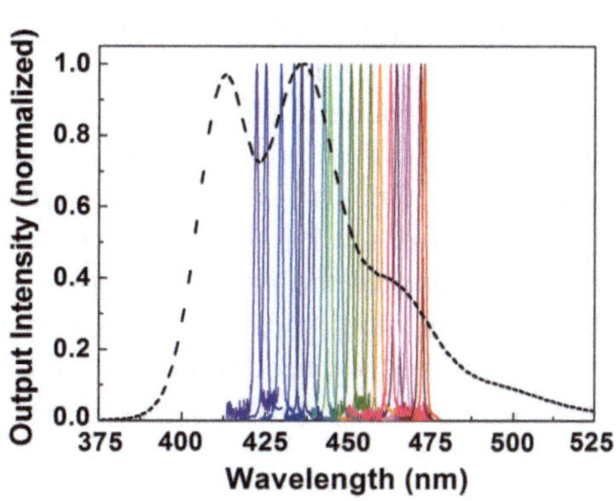

Fig. 4.14 Laser emission spectra for different grating periods of second order T3 DFB lasers with the PL spectra (*dashed line*). Lasing from 422 to 427 nm are from a 280 nm period 1D grating, 429 to 443 nm are from a 270 nm period 1D grating and 445 to 473 nm are from a 301 nm period 2D grating

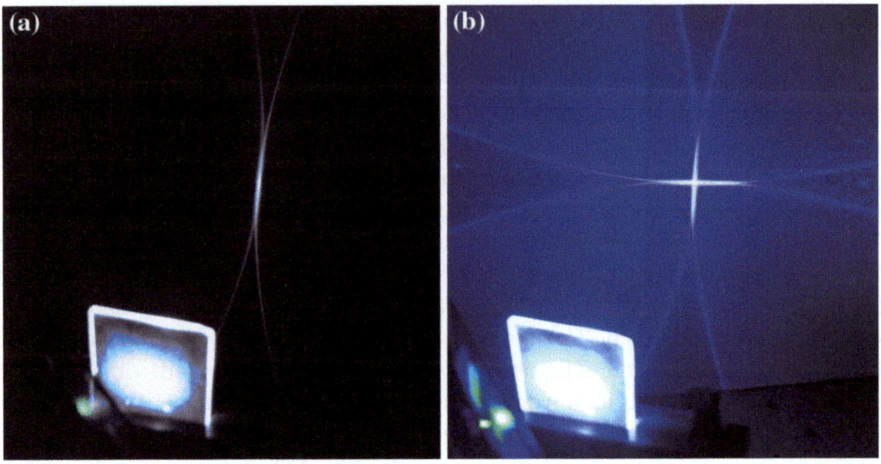

Fig. 4.15 Photographs of **a** 1D and **b** 2D T3 truxene laser beams

A final point to mention, above the laser threshold, the emission beam from the 1D laser was a fan-shaped divergent stripe, as shown in Fig. 4.15a, and parallel to the orientation of the grating grooves. In the case of the 2D laser, the emission, see Fig. 4.15b, formed a cross shape as expected for a two dimensional resonator.

4.7 Beyond 3-Armed Truxene-Oligomers

After the results presented in Sects. 4.3–4.6 were published [12, 19], Lai et al. reported detailed photophysical and laser work on a set of related monodisperse fluorene-based six-armed star-macromolecules (Tr1 to Tr4) [26]. Different from our truxene-oligomers, their molecules have two oligofluorene arms bonded to each end of the truxene core instead of one, as shown in Fig. 4.16. Very high solid-state PLQY and arm-length-dependent absorption and PL were observed. Tr3 DFB lasers showed emission that spanned the range of 421–443 nm with a lowest laser threshold of approximately 110 W/cm^2 at 437 nm with an excitation spot with a diameter of 0.018 cm.

The same synthetic chemists in Nanjing University of Posts and Telecommunications in China provided us several alternative six-arm macromolecules including the CzF_2Tr—a carbazole $C_{12}H_8N$ unit instead of a fluorene unit at the end of the arm, see Fig. 4.17. Carbazole based molecules are commonly used in OLED to improve the charge transport [27, 28]. As mentioned in Chap. 1, high charge-carrier motility is one of the essential criteria towards electrical injection lasers. However, we need to preserve the good optical gain properties at the same time as improve electrical characteristics.

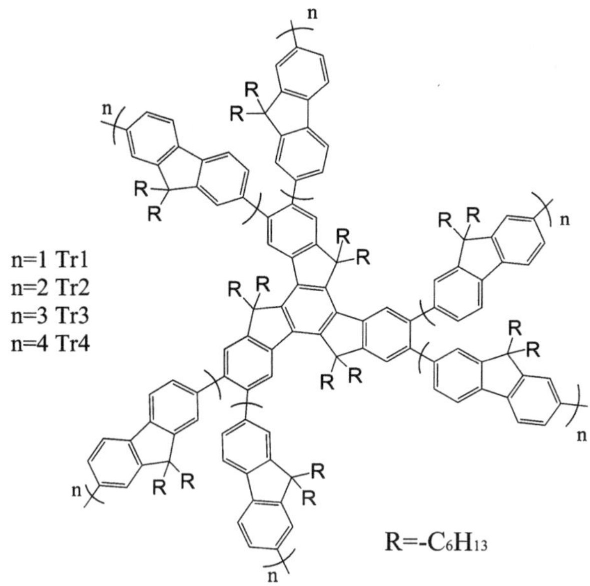

Fig. 4.16 Molecular structures of the six-armed truxene-oligomers family

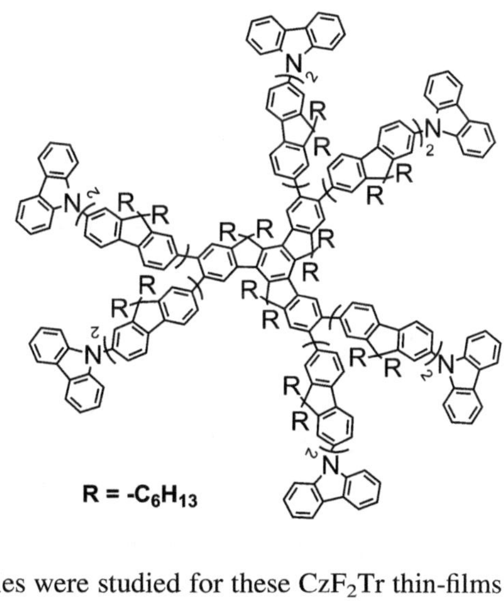

Fig. 4.17 Chemical structure of CzF_2Tr

The photophysical and gain properties were studied for these CzF_2Tr thin-films following the procedure used previously for T3 and T4. The maximum absorption occurred at 353 nm and PL peak at 430 nm. However the film PLQY was found to be 50 %. For comparison the T2 and Tr2 both have film PLQY of over 80 % [10, 26]. This lower PLQY value of CzF_2Tr was possibly due to aggregation or phase separation in the films. Figure 4.18 shows an optical microscope image of a 150 nm thick CzF_2Tr film.

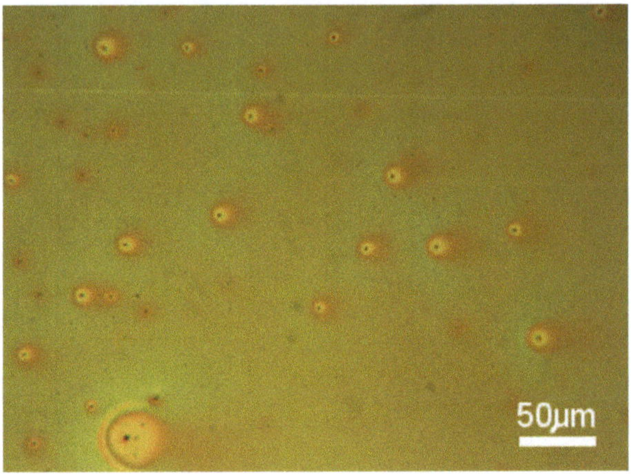

Fig. 4.18 Optical microscope image of a CzF_2Tr film spin-coated from a 30 mg/ml toluene solution on a quartz substrate

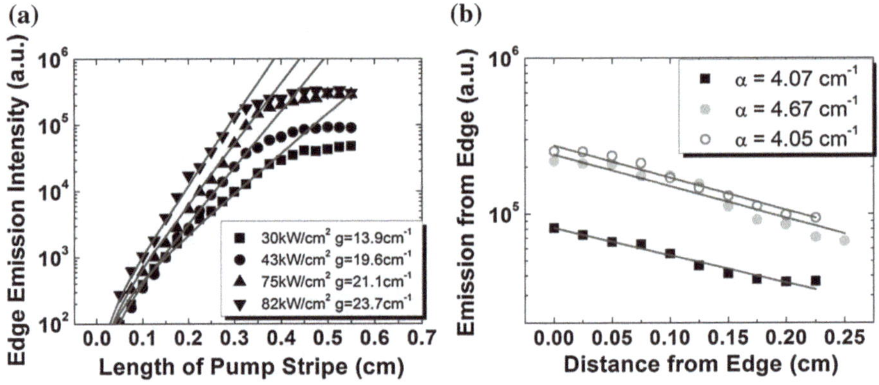

Fig. 4.19 Gain and loss measurements of CzF_2Tr waveguide

The refractive index of CzF_2Tr film at 430 nm measured with VASE was 1.80, similar to T3 and T4. The ASE peak occurred at 430 nm, with a threshold of 8.6 kW/cm^2 (34 μJ/cm^2). The maximum gain, as shown in Fig. 4.19a, was measured to be 24 cm^{-1} for a 120 nm thick air-CzF_2Tr-silica waveguide when the pump intensity was 82 kW/cm^2. The average loss coefficient was 4.5 cm^{-1}, see Fig. 4.19b, much higher than T3 and T4, along with the poor film quality it could explain the high ASE threshold.

A DFB laser was made with CzF_2Tr on a 270 nm silica grating. Laser emission occurred at 432 nm when the pump power was about 5.2 kW/cm^2 and the laser output linewidth was less than 0.5 nm.

Table 4.3 Photophysical, gain and laser performance of T3 and T4 truxenes

	T3	T4
Absorption Peak (nm)	367	375
PL Peak (nm)	414, 437, 461	420, 441, 467
PLQY (%)	86	73
ASE Threshold (kW/cm^2)	2.1	4
Optical Gain	38 cm^{-1} (58 kW/cm^2)	23 cm^{-1} (16 kW/cm^2)
Optical Loss (cm^{-1})	3.3	2.3
Lasing Threshold (W/cm^2)	515	270
Lasing Tunability (nm)	422–473	428–453

4.8 Conclusions

In this chapter detailed investigations of the photophysical characteristics and lasing performances of the T3 and T4 truxene oligomers were carried out. The materials were found to form isotropic films by spin-coating, which contributes to their low optical waveguide losses. A series of low threshold blue-emitting second order DFB lasers based on these gain media were demonstrated. Table 4.3 shows the comparisons of the photophysical properties of T3 and T4 truxene, along with the ASE and laser results.

The very low waveguide losses of these oligomer films play an essential role in achieving low lasing thresholds. These are the lowest waveguide loss coefficients reported for solution-processed neat films of organic semiconductors. Compared to the photophysical properties of T4 truxene, T3 has slightly higher solid-state PLQY and lower amplified spontaneous emission threshold. The minimum T3 lasing threshold of 515 W/cm^2 was demonstrated, although this is higher than the best T4 laser threshold, which could be due to the different resonator structures used in the lasers. Furthermore, T3 has very broad tunability of 51 nm in the deep-blue to blue region.

Unfortunately the six-arm carbazole-oligofluorene-truxene CzF_2Tr did not perform as well as hoped. Compared to T3, it had a much higher ASE threshold and loss coefficient, with lower gain (Table 4.4). Constant with these results, the DFB laser threshold was almost 10 times higher than T3 lasers.

In conclusion, this work shows that star-shaped oligofluorenes can be excellent low-threshold organic laser gain media. The PLQY and ASE performance are valuable indicators of laser behaviour. The chemical structure of the organic semiconductor plays a very important role in photophysics and waveguide characteristics.

Table 4.4 Gain properties of CzF_2Tr compared with T3

Material	ASE peak (nm)	ASE threshold(kW/cm^2)	Gain coefficient (cm^{-1})	Loss coefficient (cm^{-1})
CzF_2Tr	430	8.6	24	4.5
T3	439	2.1	38	3.3

References

1. Li, B. S., Li, J., Fu, Y. Q., & Bo, Z. S. (2004). Porphyrins with four monodisperse oligofluorene arms as efficient red light-emitting materials. *Journal of the American Chemical Society, 126*(11), 3430–3431.
2. de Bettignies, R., Nicolas, Y., Blanchard, P., Levillain, E., Nunzi, J. M., & Roncali, J. (2003). Planarized star-shaped oligothiophenes as a new class of organic semiconductors for heterojunction solar cells. *Advanced Materials, 15*(22), 1939–1943.
3. Ponomarenko, S. A., Kirchmeyer, S., Elschner, A., Huisman, B. H., Karbach, A., & Drechsler, D. (2003). Star-shaped oligothiophenes for solution-processible organic field-effect transistors. *Advanced Functional Materials, 13*(8), 591–596.
4. Kanibolotsky, A. L., Berridge, R., Skabara, P. J., Perepichka, I. F., Bradley, D. D. C., & Koeberg, M. (2004). Synthesis and properties of monodisperse oligofluorene-functionalized truxenes: Highly fluorescent star-shaped architectures. *Journal of the American Chemical Society, 126*(42), 13695–13702.
5. Gaylord, B. S., Wang, S. J., Heeger, A. J., & Bazan, G. C. (2001). Water-soluble conjugated oligomers: Effect of chain length and aggregation on photoluminescence-quenching efficiencies. *Journal of the American Chemical Society, 123*(26), 6417–6418.
6. Kanibolotsky, A. L., Perepichka, I. F., & Skabara, P. J. (2010). Star-shaped pi-conjugated oligomers and their applications in organic electronics and photonics. *Chemical Society Reviews, 39*(7), 2695–2728.
7. Gomez-Lor, B., de Frutos, O., Ceballos, P. A., Granier, T., & Echavarren, A. M. (2001). Synthesis of new C-3 h and C-3v truxene derivatives. *European Journal of Organic Chemistry, 11*, 2107–2114.
8. Greenham, N. C., Samuel, I. D. W., Hayes, G. R., Phillips, R. T., Kessener, Y. A. R. R., Moratti, S. C., et al. (1995). Measurement of absolute photoluminescence quantum efficiencies in conjugated polymers. *Chemical Physics Letters, 241*(1–2), 89–96.
9. Xia, R. D., Heliotis, G., Hou, Y. B., & Bradley, D. D. C. (2003). Fluorene-based conjugated polymer optical gain media. *Organic Electronics, 4*(2–3), 165–177.
10. Montgomery, N. A., Denis, J. C., Schumacher, S., Ruseckas, A., Skabara, P. J., Kanibolotsky, A., et al. (2011). Optical excitations in star-shaped fluorene molecules. *Journal of Physical Chemistry A, 115*(14), 2913–2919.
11. Bansal, A. K., Ruseckas, A., Shaw, P. E., & Samuel, I. D. W. (2010). Fluorescence Quenchers in Mixed Phase Polyfluorene Films. *Journal of Physical Chemistry C, 114*(41), 17864–17867.
12. Wang, Y., Tsiminis, G., Yang, Y., Ruseckas, A., Kanibolotsky, A. L., Perepichka, I. F., et al. (2010). Broadly tunable deep blue laser based on a star-shaped oligofluorene truxene. *Synthetic Met, 160*(13–14), 1397–1400.
13. Heliotis, G., Bradley, D. D. C., Turnbull, G. A., & Samuel, I. D. W. (2002). Light amplification and gain in polyfluorene waveguides. *Applied Physics Letters, 81*(3), 415–417.
14. McGehee, M. D., & Heeger, A. J. (2000). Semiconducting (conjugated) polymers as materials for solid-state lasers. *Advanced Materials, 12*(22), 1655–1668.
15. Nguyen, T. Q., Martini, I. B., Liu, J., & Schwartz, B. J. (2000). Controlling interchain interactions in conjugated polymers: The effects of chain morphology on exciton–exciton annihilation and aggregation in MEH-PPV films. *The Journal of Physical Chemistry B, 104*(2), 237–255.
16. Gupta, R., Stevenson, M., Dogariu, A., McGehee, M. D., Park, J. Y., Srdanov, V., et al. (1998). Low-threshold amplified spontaneous emission in blends of conjugated polymers. *Applied Physics Letters, 73*(24), 3492–3494.
17. Lin, H. W., Lin, C. L., Wu, C. C., Chao, T. C., & Wong, K. T. (2007). Influences of molecular orientations on stimulated emission characteristics of oligofluorene films. *Organic Electronics, 8*(2–3), 189–197.

18. Ribierre, J. C., Tsiminis, G., Richardson, S., Turnbull, G. A., Samuel, I. D. W., Barcena, H. S., et al. (2007). Amplified spontaneous emission and lasing properties of bisfluorene-cored dendrimers. *Applied Physics Letters, 91*(8), 81108–81110.
19. Tsiminis, G., Wang, Y., Shaw, P. E., Kanibolotsky, A. L., Perepichka, I. F., Dawson, M. D., et al. (2009). Low-threshold organic laser based on an oligofluorene truxene with low optical losses. *Applied Physics Letters, 94*(24), 243304–243306.
20. Heliotis, G., Xia, R. D., Turnbull, G. A., Andrew, P., Barnes, W. L., Samuel, I. D. W., et al. (2004). Emission characteristics and performance comparison of polyfluorene lasers with one- and two-dimensional distributed feedback. *Advanced Functional Materials, 14*(1), 91–97.
21. Schneider, D., Rabe, T., Riedl, T., Dobbertin, T., Werner, O., Kroger, M., et al. (2004). Deep blue widely tunable organic solid-state laser based on a spirobifluorene derivative. *Applied Physics Letters, 84*(23), 4693–4695.
22. Schneider, D., Rabe, T., Riedl, T., Dobbertin, T., KrÎger, M., Becker, E., et al. (2004). Ultrawide tuning range in doped organic solid-state lasers. *Applied Physics Letters, 85*(11), 1886–1888.
23. Woggon, T., Klinkhammer, S., & Lemmer, U. (2010). Compact spectroscopy system based on tunable organic semiconductor lasers. *Appl Phys B-Lasers O, 99*(1–2), 47–51.
24. Clark, J., & Lanzani, G. (2010). Organic photonics for communications. *Nature Photonics, 4* (7), 438–446.
25. Wallikewitz, B. H., Nikiforov, G. O., Sirringhaus, H., & Friend, R. H. (2012). A nanoimprinted, optically tuneable organic laser. *Applied Physics Letters, 100*(17), 173301–173303.
26. Lai, W. Y., Xia, R. D., He, Q. Y., Levermore, P. A., Huang, W., & Bradley, D. D. C. (2009). Enhanced solid-state luminescence and low-threshold lasing from starburst macromolecular materials. *Advanced Materials, 21*(3), 355–360.
27. v Dijken, A., Bastiaansen, J., Kiggen, N., Langeveld, B., Rothe, C., Monkman, A., et al. (2004). Carbazole compounds as host materials for triplet emitters in organic light-emitting diodes: Polymer hosts for high-efficiency light-emitting diodes. *Journal of the American Chemical Society, 126*, 7718–7727.
28. Kulkarni, A. P., Tonzola, C. J., Babel, A., & Jenekhe, S. A. (2004). Electron transport materials for organic light-emitting diodes. *Chemistry of Materials, 16*, 4556–4573.

Chapter 5
Commercial LED Pumped Organic Semiconductor Lasers

5.1 Introduction

Unlike the oligomers in Chap. 4, the number of repeat units in polymers is not one unique value. Different chains may have different lengths, which introduce general difficulties in reproducibility during the polymer synthesis. To some extent, even the same polymer can have varying optoelectronic properties depending on chain length [1]. However, polymers have played a very important role through the last decade in the development of ultra-low threshold organic semiconductor lasers (OSLs) [1–3].

As mentioned in Chap. 1, a recent breakthrough in the OSL field was the demonstration that polymer lasers could be pumped by inexpensive inorganic light-emitting diodes (LEDs), thereby providing compact, versatile laser devices with green emission [4]. Such demonstration is encouraging for the prospects of embedding OSLs in integrated devices leading to a substantial reduction in the complexity, size and expense of the optical system and so opening up new application fields. Although direct electrical pumping of OSLs is an extremely challenging problem, as discussed in Chap. 2, most of the benefits outlined above are available from simple indirect electrical pumping with inorganic LEDs. To achieve LED pumped OSL, a low threshold laser gain material is essential.

In this chapter the photophysics and laser potential of eight different batches of a green-emitting polymer, poly[2,5-bis(2′,5′-bis(2″-ethylhexyloxy)phenyl)-p-phenylene vinylene] (BBEHP-PPV) are studied. The molecular structure is shown in Fig. 5.1. In Sect. 5.2 the molecular weight and polydispersity of these batches of polymers are listed. In Sect. 5.3, photophysical properties are presented and in Sect. 5.4 the results of amplified spontaneous emission from thin-film waveguides of all the different batches are discussed. Their optical properties are compared against molecular weights in Sect. 5.5 and the optical constants from some batches are studied. In Sect. 5.6 distributed feedback lasers made from the low ASE threshold batches are characterised and the optimization of the lasing threshold with several different approaches are discussed in detail.

Y. Wang, *Low Threshold Organic Semiconductor Lasers*, Springer Theses,
DOI: 10.1007/978-3-319-01267-4_5, © Springer International Publishing Switzerland 2014

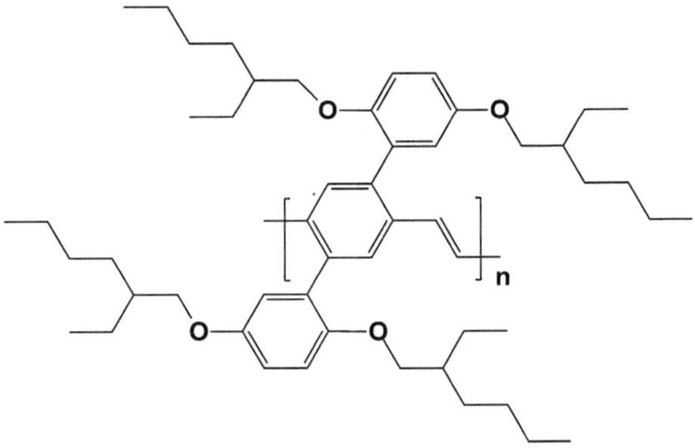

Fig. 5.1 Molecular structure of BBEHP-PPV polymer

To explore the potential of using low-cost LEDs to pump the organic lasers, power intensities of several commercially available LEDs are investigated and particularly high-power LEDs are examined in Sect. 5.7 with a current-pulse driver. Section 5.8 presents full characterization of an InGaN LED pumped BBEHP-PPV laser, followed by the conclusions in Sect. 5.9.

5.2 Batches of Polymer BBEHP-PPV

The molecular weight of a synthetic polymer does not have a single value due to the chain-length variation. It is common therefore to calculate the average molecular weight of a polymer, i.e. a distribution of molecular weights. This distribution depends on the way the polymer is produced [5, 6].

During the HYPIX project a total of 8 batches of BBEHP-PPV were synthesised by Dr. Alexander Kanibolotsky and Prof. Peter Skabara at the University of Strathclyde. These are listed in Table 5.1. Initially 3 batches were studied: AK342P was the unrefined BBEHP-PPV polymer; AK350DCM was obtained by an additional purification process—soxhlet extraction with dichloromethane (DCM, with the formula CH_2Cl_2) as the solvent, giving a higher molecular weight (MW) fraction; and AK350Me2CO was produced from a further acetone soxhlet extraction, which is expected to appear with lower MW.

As mentioned above, the difficulties in reproducibility of polymers introduce the batch-to-batch differences in these BBEHP-PPV polymers, resulting in the variation in a photophysical and lasing properties which will be discussed in detail in the following sections. Nevertheless, from the initial 3 batches we learned how the materials' optical properties in which we are most interested vary with the M_W or the polydispersity index (PDI) values.

After supplying the optical property information back to the chemists, another five batches were subsequently supplied: AK391DCM1B, with high M_W; AK391DCM1F, with extremely high M_W; AK391DCM2, and later in HYPIX, we

Table 5.1 GPC data acquired by multi-angle light scattering size exclusion chromatography

Batch name	Solvent	m Injected	$M_{N(g/mol)}$	$M_{w(g/mol)}$	PDI
AK342P	THF	0.876	54,790	582,800	10.638
AK350DCM	DCM	2.314	91,310	455,300	4.987
AK350Me2CO	Acetone	2.218	28,270	28,430	1.006
AK391DCM2	DCM	2.1	96,930	205,300	2.118
AK391DCMB	DCM	2.2	74,540	341,100	4.576
AK391DCMF	DCM	2.48	143,700	595,100	4.142
AK408DCM	DCM	2.56	79,480	543,400	6.838
AK408R	THF	2.58	65,000	472,800	7.27

received the batch AK408DCM and AK408R (re-precipitated from tetrahydrofuran (THF) with the formula $(CH_2)_4O$). Table 5.1 presents the Gel permeation chromatography (GPC) data obtained by the chemistry group at the University of Strathclyde, including the number average molecular weight (M_N), the weight average molecular weight (M_w) and the polydispersity index (PDI).

$$\mathrm{PDI} = \frac{M_N}{M_W} \tag{5.1}$$

Generally speaking, the polymers after soxhlet extraction with DCM have lower polydispersity compared with THF extraction fraction, and after acetone extraction the polymers have much lower weight-average molecular weight, M_W, than the polymers from the THF or DCM extraction.

5.3 Photophysical Studies

Absorption and photoluminescence spectra were measured from thin films (spin-coated from 10 mg/ml chlorobenzene solutions on silica substrates). Films were excited at 430 nm, giving two distinguishable emission peaks at around 495 and 530 nm with a shoulder around 570 nm, as shown in Fig. 5.2.

From Tables 5.1 and 5.2, we can see that the different batches of polymers have very similar absorption and PL peak-wavelengths, except that the very low M_W batch, AK350Me2CO, has a-few-nm blue-shift.

Solid-state PLQY were measured using a Hamamatsu absolute PLQY measurement system, as described in Chap. 3. It is important to clarify that there is no chain aggregation in the spin-coated polymer film so the PLQY in both powder and neat film were measured to clarify the absence of film aggregation. A quartz cuvette containing the powder was placed inside the integrating sphere when measuring the absolute PLQY in powder. The excitation wavelength in these measurements was selected by the monochromator, to be 450 nm. The films made from most batches had very high PLQY, above 80 %. The results presented in Table 5.3 suggest that there is no strong aggregation affecting the neat BBEHP-PPV films.

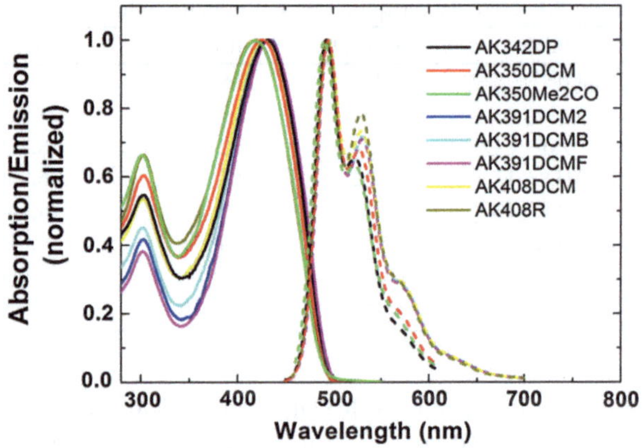

Fig. 5.2 Absorption (*solid line*) and photoluminescence (*dashed line*) spectra of all different batches of BBEHP-PPV

Table 5.2 Absorption and PL peaks of various BBEHP-PPV

	AK350Me2CO $M_W = 28{,}430$	AK391DCMB $M_W = 341{,}300$	AK350DCM $M_W = 455{,}300$	AK391DCMF $M_W = 595{,}100$
Abs. peak (nm)	419	431	431	432
PL peaks (nm)	491, 524, 563	495, 530, 569	495, 528, 568	495, 530, 570

Table 5.3 PLQY of different batches of BBEHP-PPV polymer

Batch name	Powder PLQY (%)	Film PLQY (%)
AK342P	74	82
AK350DCM	76	77
AK350Me2CO	79	82
AK391DCM2	74	81
AK391DCMB	78	84
AK391DCMF	72	84
AK408DCM	84	86
AK408R	86	88

PLQY measured in this method has uncertainty of ±2 %

The excited-state lifetime of BBEHP-PPV (AK350DCM) thin-films was measured to be 590 ± 20 ps. Combined with the film PLQY value, the radiative lifetime was calculated to be approximately 760 ps.

5.4 ASE Studies of BBEHP-PPV Films

For ASE measurements, film samples of around 200 nm thickness, prepared from 15 to 25 mg/ml chlorobenzene solutions, were placed in a vacuum chamber under a pressure of approximately 10^{-4} mbar. The excitation light was generated by an OPO with 4 ns output pulses at a repetition rate of 20 Hz, as described in Chap. 3.

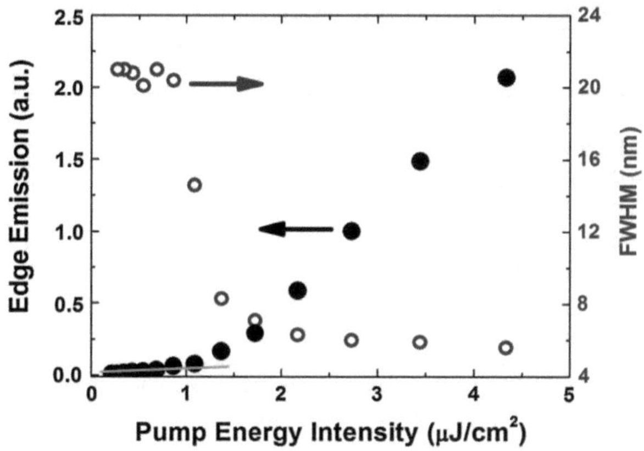

Fig. 5.3 ASE threshold of a AK350DCM waveguide; polymer film thickness is 210 nm

The pump wavelength was selected to be 430 nm, close to the peak absorption of BBEHP-PPV films. A cylindrical lens was used to shape the excitation beam into a narrow stripe with a width of approximately 460 μm on the surface of the film. The length of the pump stripe was 4.7 mm, which resulted in an excitation area of 0.0216 cm^2. The edge emission from the film was collected and recorded with a fibre-coupled grating spectrograph with a CCD detector.

Figure 5.3 shows for AK350DCM, the full-width at half-maximum linewidth (FWHM) of the edge-emission spectrum and relative output intensity as a function of the pump intensity. The FWHM dropped rapidly from 20 to 8 nm when pumping above a threshold intensity of 1.1 $\mu J/cm^2$, equivalent to 270 W/cm^2. The peak wavelength of the line-narrowed ASE appeared at 534 nm.

By measuring the ASE intensity as a function of pump stripe length at different pump intensities, the maximum net gain coefficient was found to be 19 cm^{-1} for a pump peak intensity of 3 kW/cm^2. At higher pump intensities, the gain coefficient no longer increased with the rising pump power indicating gain saturation in the polymer waveguide. The light propagating inside the waveguide experienced a very low loss, 0.83 cm^{-1}, measured using the same method as described in Chap. 3.

Table 5.4 summarises the ASE results from all the batches of BBEHP-PPV. Most batches give an ASE threshold in the range of 200 to 300 W/cm^2. It is worth noting that the two highest ASE thresholds were from the lowest and the highest molecular weight batches this will be discussed in more details in the next section.

5.5 Effect of Molecular Weight

The variation in the ASE thresholds can be related to the different average molecular weights of the different batches. As mentioned in Sect. 5.2, the synthetic polymers have a distribution of molecular weights, which is due to the different lengths of the polymer chains. Figure 5.4 plots the ASE threshold and PLQY as a

Table 5.4 Summary of ASE measurement of different batch BBEHP-PPV thin-film waveguides

Batch name	ASE wavelength (nm)	FWHM (nm)	ASE threshold (W/cm^2)
AK342P	535	6	300
AK350DCM	534	6	270
AK350Me2CO[a]	530	6	490
AK391DCM2	535	9	220
AK391DCMB	535	8	170
AK391DCMF	532	7	640
AK408DCM	532	9	280
AK408R	530	8	270

[a] AK350Me2CO was dissolved in hexane rather than chlorobenzene, as its chlorobenzene solution did not spin good-quality film

function of M_W, M_N and *PDI* for the 8 batches of BBEHP-PPV. Excluding the lowest and highest molecular weight batches, AK350Me2CO and AK391DCMF, respectively, the ASE thresholds and PLQYs were similar for the other 6 batches, however smaller polydispersity polymers had slightly lower ASE threshold.

Figure 5.4 (a) and (b) show that when the M_W is between 200,000 and 600,000 and the M_N is between 50,000 and 100,000, the ASE thresholds are in the range of 170 to 300 W/cm^2. With a molecular weight too high or too low both resulted in a much higher ASE threshold. The highest molecular weight batch AK391DCMF, shown in the photos of Fig. 5.5, was fibrotic in form, in contrast with the powders of all other batches. The reason for this is not clear but the different morphology could result in more light-scattering loss in the thin-film waveguide. The waveguide loss coefficient of an AK391DCMF film was measured to be 2.4 cm^{-1}. The ASE threshold of this batch was particularly high, 640 W/cm^2.

Excluding the lowest and highest M_W batches, Fig. 5.4 (c) indicates that in polymer synthesis a low polydispersity could help to reduce the ASE threshold while the polymer's molecular weight sits in the appropriate range. The batch AK350Me2CO had the smallest polydispersity index (PDI) among these eight batches, it also had by far the lowest average molecular weight, M_N and M_W. This is to say, the majority of polymer chains in this batch had relatively short lengths. It has been reported that a reduction in both the absorption coefficient and PL efficiency are expected with a decreased chain-length [7, 8]. However, in Fig. 5.4 (d), we do not see a strong dependence between the film PLQY and the molecular weight. On the other hand, this particular batch of BBEHP-PPV polymer formed films with very poor surface quality as observed by optical microscopy, which could result in extra light scattering and hence increase in the waveguide loss. This could explain the relatively high ASE threshold of the batch AK350Me2CO, 490 W/cm^2.

The lowest threshold, 170 W/cm^2 was obtained from the batch AK391DCMB ($M_N = 74{,}540$, $M_W = 341{,}100$), which was more than a magnitude lower than the truxene-oligomers in Chap. 4.

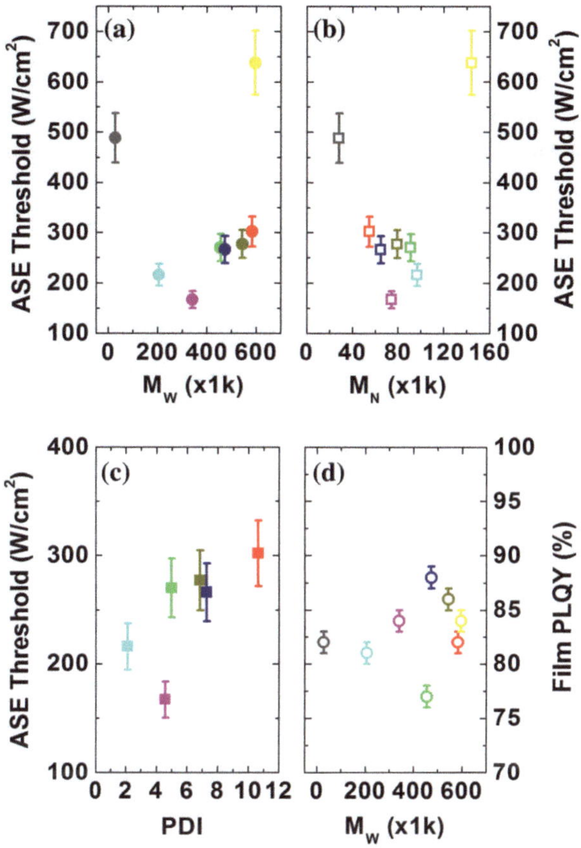

Fig. 5.4 ASE threshold dependence of **a** weight-average molecular weights (M_W), **b** number-average molecular weights (M_N) and **c** polydispersity index (PDI) of different batches of BBEHP-PPV polymers, AK342P (*red*), AK350DCM (*green*), AK350Me2CO (*gray*), AK391DCM2 (*cyan*), AK391DCMB (*magenta*), AK391DCMF (*yellow*), AK408DCM (*dark yellow*) and AK408R (*navy*); **d** the film PLQY dependence of M_W

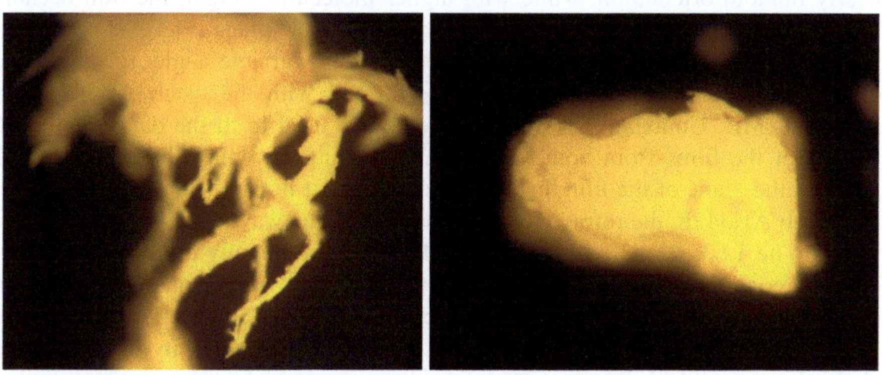

Fig. 5.5 Optical microscopy images of polymer (*left*) AK391DCMF and (*right*) AK391DCMB; the AK391DCMF is much more fibre-like compared to the other batches

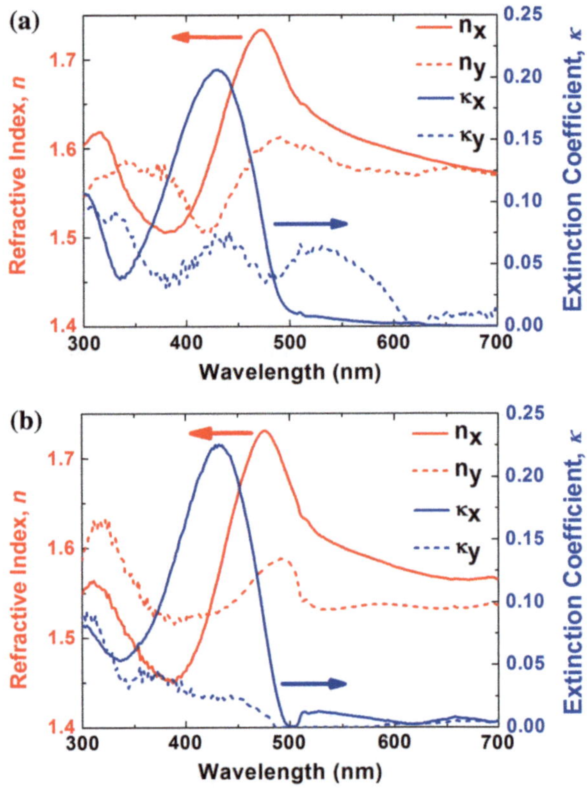

Fig. 5.6 Refractive index (*red*) and extinction coefficient (*blue*) in-plane (*solid line*) and out-of-plane (*dashed line*) of the BBEHP-PPV thin-films, **a** AK342P and **b** AK391DCMB

The investigation of molecular-weight dependent optical properties itself is a very interesting area. Koynov et al. reported comprehensive refractive index studies of MEH-PPV, 8 with a very broad range of molecular weight, 10,000–1,600,000. They found that the film morphology depends significantly on the molecular weight when it was less than 400,000. With low molecular weights, the polymer chains had nearly random orientation, while with higher molecular weight, the film became more anisotropic with chains preferentially lying in the plane [9].

Due to the limited quantity of the materials, variable-angle spectroscopic ellipsometry measurements were only performed with the batch AK342P and AK391DCMB. Uniaxial anisotropic models were used to fit the data. Figure 5.6 show that the films from both batches are birefringent, with a higher refractive index in the plane of the film than perpendicular to it in the visible spectral range. For batch AK342P, the refractive index in-plane at the wavelength of 530 nm was found to be 1.63, relatively low compared to other commonly used green-yellow-emitting conjugated polymers, for example Super Yellow mentioned in Sect. 3.3.4 (n_o = 1.93 at 530 nm). This lower refractive index is probably the result of the bulky side-groups in BBEHP-PPV which leads to a lower density of conjugated chains in the solid film.

5.6 Low Threshold BBEHP-PPV DFB Laser

As introduced in Chap. 1, one of the main goals of the HYPIX project is to demonstrate low threshold organic semiconductor lasers, which potentially could use light-emitting diodes (LEDs) as the pump source. The extremely low ASE threshold of BBEHP-PPV waveguide was a good start, and the measurement of the ASE threshold across all the batches of BBEHP-PPV, along with the optical constant measurements, would help to identify a candidate for a low threshold polymer laser. In this section, surface-emitting DFB lasers made with this polymer are investigated in terms of optimizing laser threshold.

5.6.1 Optimize the Lasing Wavelength

When the period of a DFB laser resonator is fixed, the effective refractive index, n_{eff}, in the equation $m\lambda_{\text{Bragg}} = 2n_{\text{eff}}\Lambda$, is the key parameter determining the Bragg wavelength. If the lasing wavelength is set close to the ASE peak, the DFB laser is expected to have a lower threshold than at other wavelengths. Thus a lasing threshold optimization can be achieved by tuning the lasing wavelength, i.e. varying the effective refractive index of the waveguide, by changing the film thickness.

It is important to note that the film thickness also affects waveguiding wavelength in the film. For thick films, ASE occurs at a optimum wavelength which has lowest threshold; if the films are made too thin, the optimum wavelength is no longer waveguided, then the actual ASE peak wavelength follows the cut-off wavelength of the waveguide. Figure 5.7 illustrated this by showing how the film thickness affects the ASE wavelength and threshold. When the AK350Me2CO polymer film was too thin, it resulted in the guided mode in the polymer film being too close to the cut-off wavelength (such as point A in Fig. 5.7). The ASE threshold was reduced by having a thicker polymer film (such as points B and C).

Ideally we would design the polymer laser to ensure the lasing occurs at the peak wavelength of ASE, where the gain is maximal. Figure 5.8 shows how the laser threshold varies with the lasing wavelength for the batch AK342P. A one-dimensional holographic grating with period of 355 ± 5 nm was used here. The wavelength of the lowest threshold laser was at around 535 nm.

Another consideration for low threshold laser operation is that a single transverse mode output is preferable [4, 10]. This means that the polymer film thickness should not be so thick that it would support multiple transverse modes. Figure 5.9 shows the cut-off film thickness for different wavelengths for the fundamental TE and TM modes, calculated with the refractive index of the batch AK342P. This provided a guideline for the BBEHP-PPV polymer film thickness in later laser fabrication.

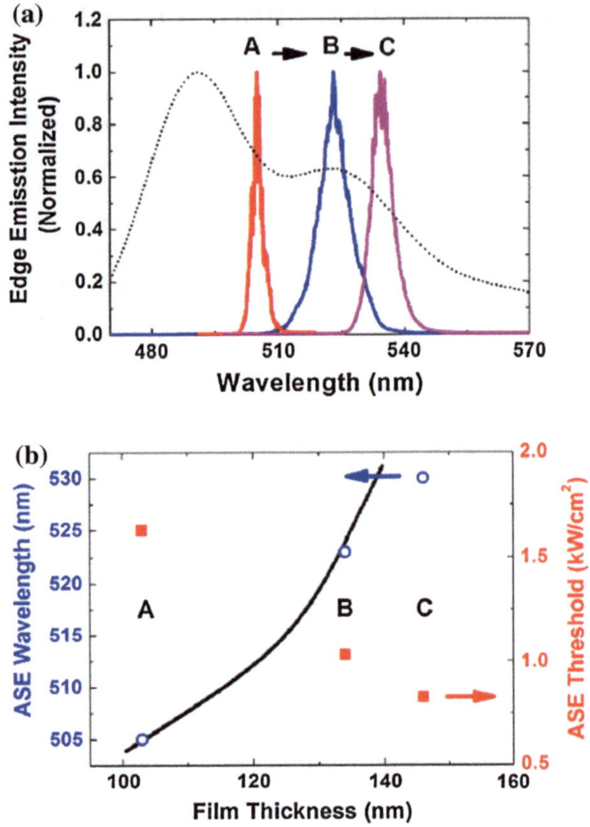

Fig. 5.7 **a** ASE spectra from AK350Me2CO films with different thicknesses. A, 103 nm; B, 134 nm; C, 146 nm; the *dotted line* on the background is thin-film PL spectrum; **b** ASE peak wavelength and threshold as function of film thickness, the *solid line* indicates the estimated cut-off wavelength of waveguided ASE mode in the BBEHP-PPV films

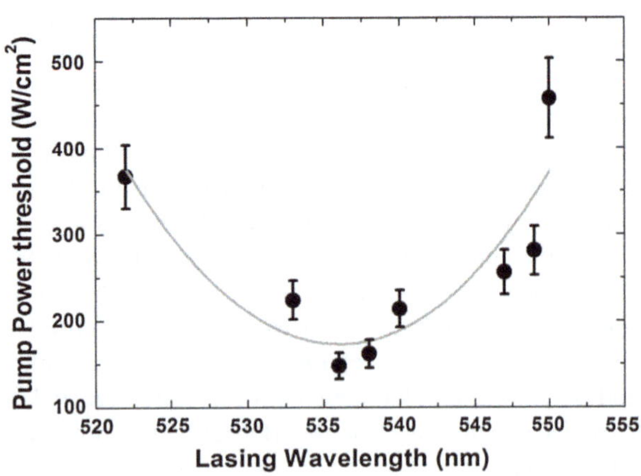

Fig. 5.8 Second order DFB lasers are made with a holographic grating with corrugation period of 355 ± 5 nm; the laser threshold as a function of output wavelength; the *solid line* is a guide for the eye

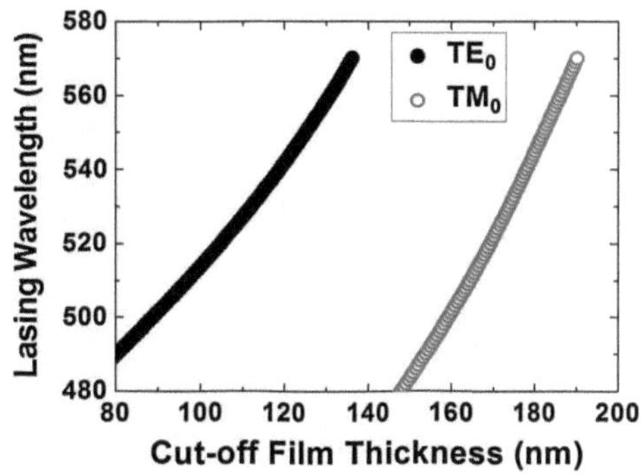

Fig. 5.9 Calculated cut-off film thickness for different TE_0 and TM_0 lasing wavelength of the batch AK342P

5.6.2 Choice of Laser Fabrication Method

Second order DFB resonators in this chapter were fabricated using two methods, solvent-assisted micro-moulding (SAMIM) on the film surface and holographic grating etched into silica substrate. The corrugation periods in both resonators were same, i.e. 355 nm, to provide the optical feedback at a wavelength around 535 nm. The pump source used here was 450 nm from the OPO (4 ns, 20 Hz). From Fig. 5.10, we can see that at the same lasing wavelength the laser made with the holographic grating has lower threshold.

5.6.3 Optimization of Laser Encapsulation

Two techniques were used to encapsulate the polymer lasers, the first one was spin-coating a thin layer of CYTOP (Asahi Glass, Japan), a chemically robust and optically transparent amorphous fluoropolymer, directly on to the polymer film, and the second was to use Norland UV-curable adhesive to glue a coverslip over the device in order to "seal" the top of the laser.

The laser threshold with CYTOP encapsulation was measured after being left for different amounts of time in air; the results indicated that the encapsulated laser threshold remained the same for ~48 h. The UV-curable adhesive encapsulation method could extend the laser lifetime for at least a few months. The BBEHP-PPV laser performance with these two different encapsulation techniques were tested and the results are compared in Table 5.5.

From Table 5.5, we can see that the lasing mode of the glue encapsulated laser red-shifts by 11 nm and results in a substantial increase of lasing threshold, by 42 %, while the CYTOP layer (thickness approximately 1.5 μm) has little impact on the laser threshold, with only 2 nm red-shift of the lasing wavelength. The red–shifts were due to the change of effective refractive index that induced by the additional encapsulation layer.

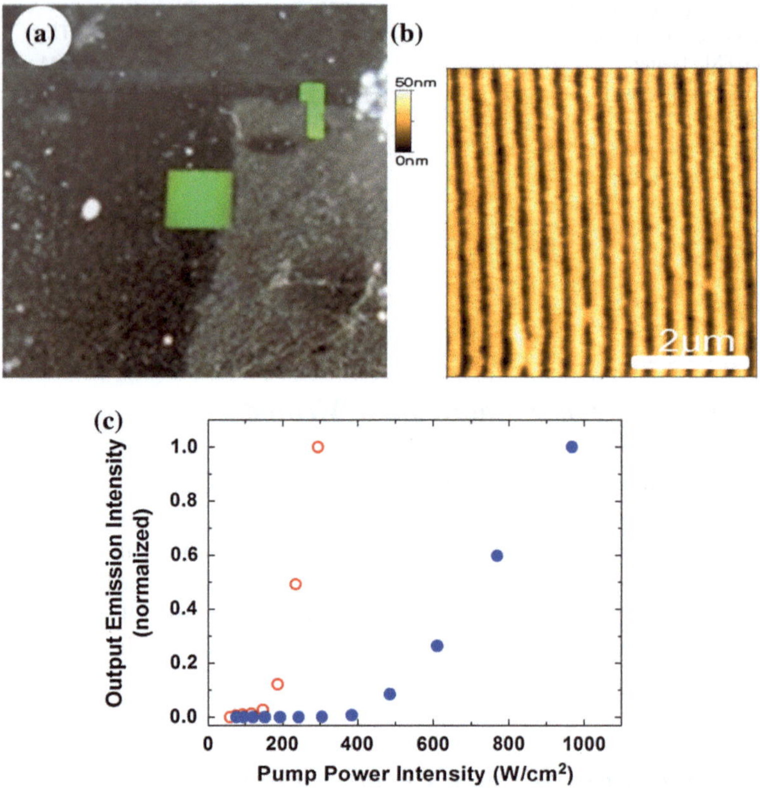

Fig. 5.10 **a** Photograph of SAMIM copolymer stamp with 1D 355 nm gratings in the middle, green light diffraction is observed with the naked eye; **b** AFM image of a corrugated polymer film using the SAMIM stamp; **c** laser threshold at 537 nm made by two different techniques, holographic grating (*open circle*) and SAMIM (*closed circle*), the pump thresholds are 150 and 385 W/cm^2, respectively

Table 5.5 Encapsulated laser performance comparison

Threshold before encapsulation	Encapsulation method	Threshold after encapsulation	Lasing wavelength shift (nm)
167 W/cm^2 @538 nm	Glue + coverslip	238 W/cm^2 @ 549 nm	11
295 W/cm^2 @540 nm	CYTOP	317 W/cm^2 @ 542 nm	2

Figure 5.11 plots the intensity profile of the fundamental TE modes for a 170 nm thick BBEHP-PPV film on silica substrate (refractive index 1.46), with a CYTOP capping layer (1.5 μm). The refractive index of CYTOP is around 1.35 over the visible spectrum. Without the capping layer, the effective refractive index

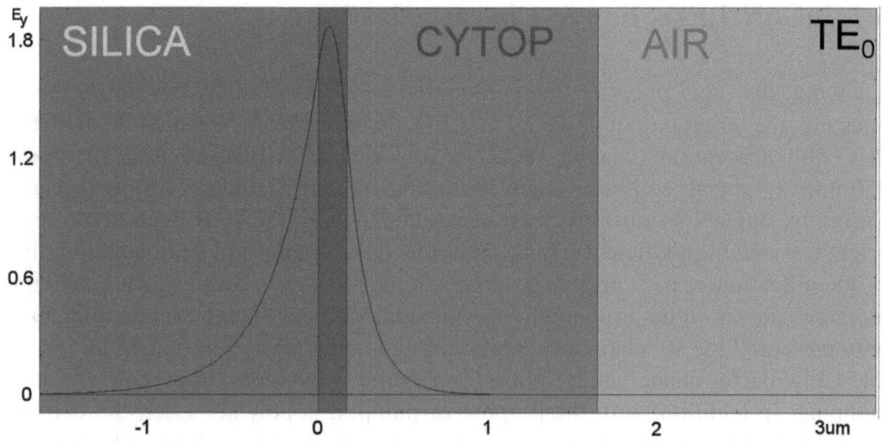

Fig. 5.11 Fundamental TE mode intensity profile in a silica-polymer-CYTOP-air structure; plotted using an online mode solves: http://wwwhome.math.utwente.nl/~hammerm/oms.html

of the thin-film waveguide for TE_0 mode at 535 nm was calculated to be 1.48; With CYTOP, the effective refractive index was calculated to be 1.50. We can see from Fig. 5.11 that the TE_0 mode is well confined within the gain medium. Strong coupling arising from the good optical confinement of the laser mode would potentially lead to low laser threshold [11].

With all the effort above, low threshold surface-emitting DFB lasers were fabricated by spin-coating a thin film of BBEHP-PPV (AK391DCMB) onto a 1D silica grating with periods of 355 ± 5 nm, encapsulated with CYTOP. The feedback wavelength was at 533 nm, with a 170 ± 20 nm thick film. Using 4 ns (FWHM) pulses at 450 nm from the OPO as the optical excitation (pump beam 0.97 mm in diameter), the AK391DCMB laser threshold was measured to be as low as 118 W/cm^2, see Fig. 5.12.

Fig. 5.12 Surface-emitting BBEHP-PPV laser threshold

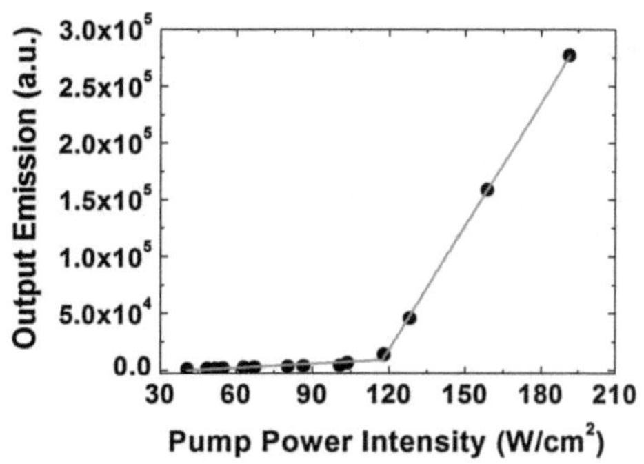

5.7 InGaN LEDs Power Characterization

Knowing that the polymer laser requires a-few-hundred W/cm^2 pump peak intensity, the target output power of an LED was then clear. To match the BBEHP-PPV film absorption (peaked at 427 nm), we chose LEDs emitting at around 450 nm, a typical emission wavelength for Indium Gallium Nitride (InGaN). Currently, InGaN is also the best candidate material for fabrication of high-brightness and high-efficient LEDs, driven by developments in solid-state lighting.

Identification of high brightness commercial LEDs was based on the maximum luminous power value provided by the manufacturers, although the manufacturers only presented the standard specification tested under CW operation while instead the LED performance under pulsed operation was more important for laser pumping. In addition, with the purpose of pumping a polymer laser, the compatibility of the LED and the laser was also considered. A few modules of high power LEDs were purchased and tested. Here the Philips LUMILED LUXEON Rebel LED is taken as an example to explain the characterization process, as it was in the end selected as the best candidate for pumping the polymer laser.

The emission colour of the LED chosen was royal-blue (LXML-PR01-0500), with a peak emission wavelength of 448 nm, see Fig. 5.13. The light emitting area of the LED is 1.3 mm by 1.3 mm. The power output of the LED is typically 875 mW at 700 mA under CW-operation mode as specified, which is approximately 50 W/cm^2. In order to pump the polymer laser, the InGaN LED was driven with nanosecond pulses generated by a compact laser diode driver (module PCO-7110-120-15, Directed Energy Inc.), designed to provide high current (up to 120 A) pulses. The pulse duration and peak current depend on the high voltage supplied to the circuit. The trigger signal for the driver was supplied separately by a function generator, which also set the repetition frequency of the LED pulses. The output pulse energy from the LED was measured with a calibrated energy meter for a range of driving currents. The optical pulse duration was monitored with a fast photodiode (HSP-V2, Standa) on a 500 MHz bandwidth oscilloscope. The fast photodiode has a bandwidth of more than 2 GHz with rising and falling

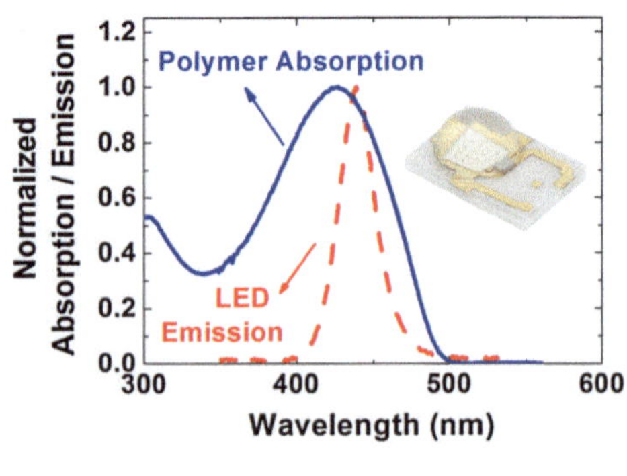

Fig. 5.13 Normalized absorption spectrum overlap with the LED emission spectrum; Inset, photograph of a LUXEON Rebel LED before the plastic lens removed

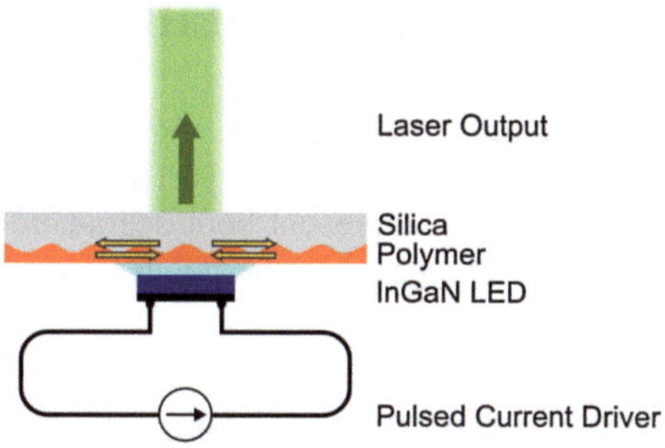

Fig. 5.14 Schematic drawing of a LED pumped polymer laser

time less than 150 ps. Figure 5.14 presents a schematic drawing of the LED pumped polymer laser geometry. To be able to access the maximum peak intensity from the LED surface, the plastic lens was removed without damaging the light-emitting chip, so that the polymer laser could be placed as close as possible to the LED chip, thus minimizing the effect of light divergence from the LED.

The energy density from the LED as a function of driving current at 20 Hz repetition rate is shown in Fig. 5.15, as well as the corresponding FWHM of the optical pulse duration. As the peak driving current increased to 40 A, the energy density at the light-emitting surface could reach as high as 37 μJ/cm^2, i.e. 840 W/cm^2 with pulse duration of 45 ± 2 ns. The pulsed current damage threshold was beyond 50 A, which slightly varied between LEDs. The maximum peak intensity of these LEDs was approximately 1,000 W/cm^2, 20-times higher compared to the output intensity under CW operation. With such high intensities, there was the prospect that the BBEHP-PPV lasers could be pumped by these blue LEDs.

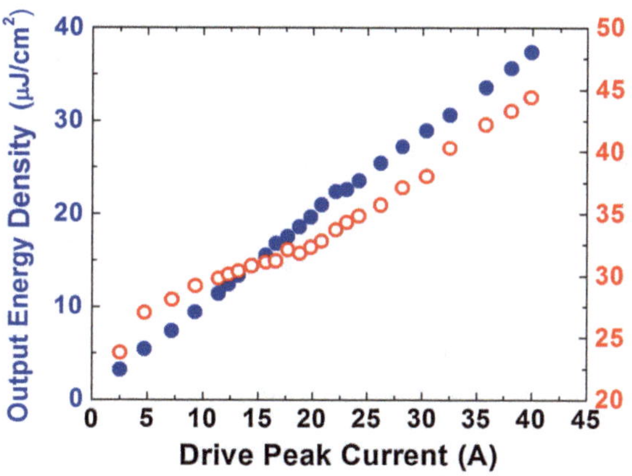

Fig. 5.15 Output energy density (*closed circles*) from a pulsed LUXEON LED (at 20 Hz repetition rate) as a function of driving current and the corresponding optical pulse duration (FWHM) (*open circles*)

5.8 LED Pumped BBEHP-PPV Lasers

Second order DFB lasers using the LUXEON LED as the pump source were investigated in detail. The BBEHP-PPV lasers were encapsulated with the CYTOP capping layer to prevent both degradation of the polymer in the air and possible physical damage when the polymer was brought into contact with the LED.

5.8.1 *Laser Emission Spectra Study*

The shaded areas in Fig. 5.16 (a) indicate three distinct regions during the laser build-up process. At low levels of driving current (region 1), corresponding to pump intensity below 500 W/cm^2, a very pronounced Bragg dip was evident at 532.8 nm within an 8 nm (FWHM) gain envelop. With the driving current increased to above 15.7 A, corresponding to a peak intensity of 500 W/cm^2, the

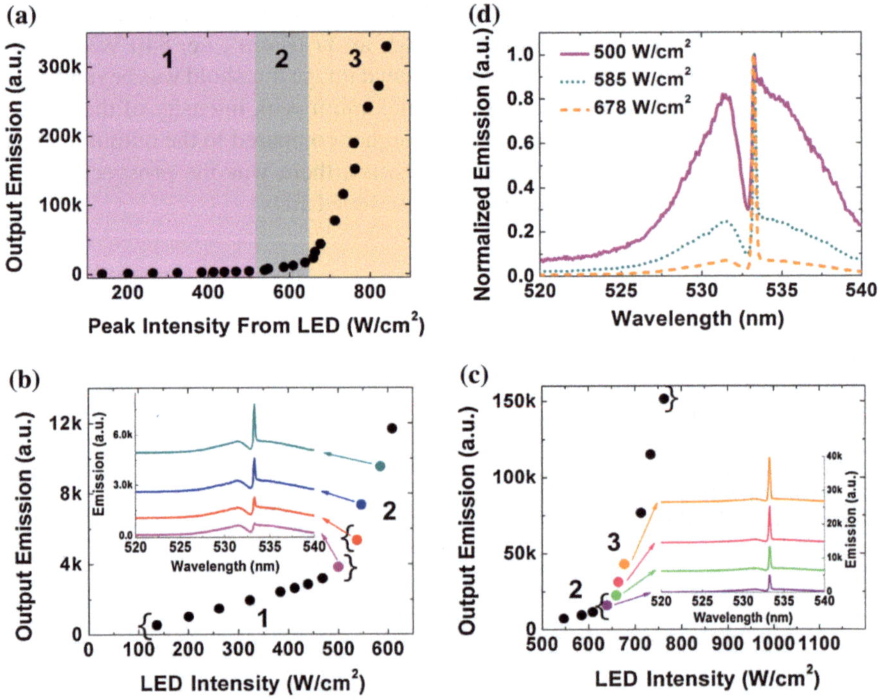

Fig. 5.16 **a** Polymer output emission intensity at the wavelength of 533 nm as a function of LED peak intensity; **b** and **c** Lasing intensity as a function of LED peak intensity in different regions. Inset: laser spectra collected at different pump intensity levels; **d** Overlap of normalized emission spectra at three different LED power intensities, (*solid line*) 500 W/cm^2, (*dotted line*) 585 W/cm^2 and (*dashed line*) 678 W/cm^2

laser started to build up at the wavelength of 533.2 nm, spectrally located at the band-edge of the Bragg dip, as shown in Fig. 5.16 (b). The linewidth of the peak at 533.2 nm narrowed to <0.3 nm as the peak current increased further in region 2, while at the same time the output intensity at the peak wavelength increased significantly, appearing as a sudden change in the slope of the output intensity. A second abrupt slope change in the output versus input plot occurred when the LED peak intensity was further increased to 639 W/cm^2. The narrow lasing peak at 533.2 nm dominated the emission spectrum in region 3, as shown in Fig. 5.16 (c). The laser linewidth with a pump intensity of 678 W/cm^2 was measured to be 0.2 nm, with the resolution of the spectrograph used in these measurements being 0.1 nm at the appropriate wavelength.

To explain this, in region 1 for pump intensity below lasing threshold, the output spectrum mainly consisted of waveguided PL that scattered out of the film at the normal angle to the substrate, i.e. spontaneous emission. The Bragg dip in the spectrum was due to an inhibited propagation of the waveguided PL in the stop-band of the grating [12]. In region 2 when the LED intensity increased, laser oscillation started since the gain exceeded the loss in the gain medium. However, the spontaneous emission was still strongly evident in the spectrum because the LED pulse duration was longer than 30 ns with a slow rising edge (>10 ns), while the lasing only happened in a few nanoseconds. The onset of lasing was the reason for the change in the output slope at 533 nm. In region 3 with even higher excitation intensity, the laser turned on earlier in the pump pulse and finally the stimulated emission dominated the gain, resulting in a second slope change. This will be discussed more in the next section. The spectra from these three pump regions are directly compared in Fig. 5.16 (d). It is obvious that the spectral linewidth of the laser emission peak from regions 2 and 3, with and without the Bragg scattering shoulder, remains the same. This is one of the criteria of identify lasing action [13].

This 'double-threshold' phenomenon was not observed when using the OPO as pump source due to its much shorter pulse-duration and steeper rising edge compared to the LED pulses. The reason for this is that the faster rising time of the OPO pulse allows the polymer laser to build up much quicker.

The reason that the LED pumped polymer laser threshold (500 W/cm^2) was higher than that observed with OPO pumping (120 W/cm^2) was possibly that non-radiative triplet excitons built up during the >30 ns pump pulses from the LED creating extra loss mechanism. The long pulses would induce heating in the polymer film which might also increase the polymer laser threshold.

5.8.2 Laser Pulse Dynamics Study

The dynamics of polymer laser output pulses along with the LED pulses were recorded at different levels of driving current using a fast-photodiode. In region 1, the "below-threshold laser" (spontaneous emission) output pulse duration was as long as the LED pulse as expected, i.e. the “laser” switched on as soon as the LED turned on; In region 2, as the LED peak intensity increasing, the laser pulse became longer (increasing from $\sim$3 to $\sim$8 ns), and the polymer laser switched on faster as the LED intensity got higher, as shown in Fig. 5.17 (b); In region 3, the polymer laser build-up time became constant ($\sim$5 ns) and the laser pulse duration also stabilised around 9–11 ns, possibly because of the triplet building-up during such long pumping pulses which terminated laser action. Figure 5.17 (c) clearly shows in region 2 the output emission is a combination of spontaneous emission and laser.

It is worth noting that although the laser terminated about 25 ns before the LED pulse reduced to its half maximum, the laser pulses were still much longer than the BBEHP-PPV polymer's excited state lifetime ($\sim$600 ps).

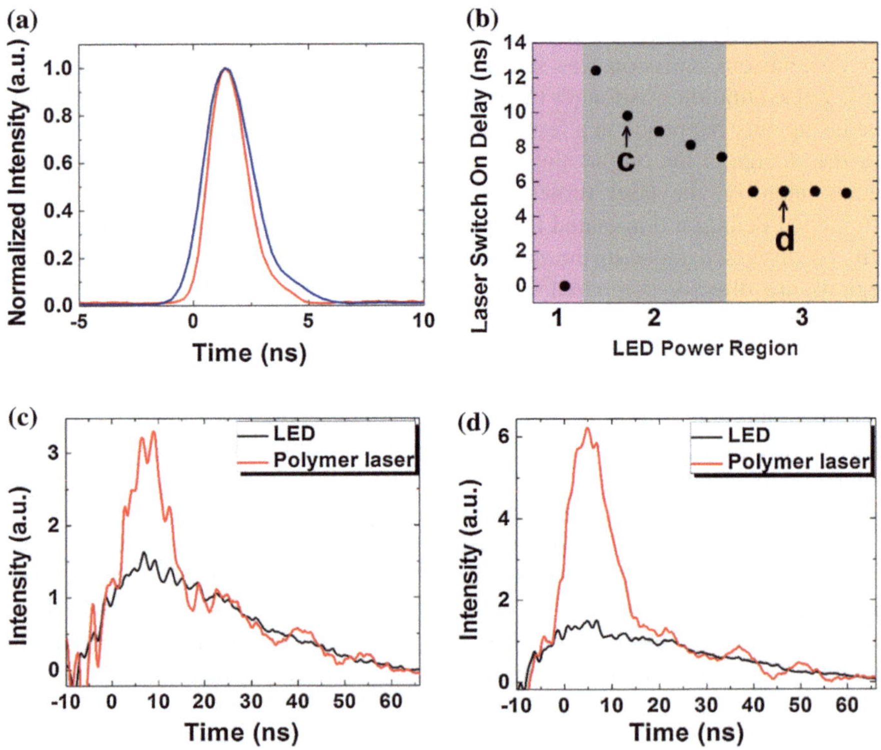

Fig. 5.17 **a** Dynamics of laser output pulses from OPO (*blue*) and polymer above threshold (*red*); **b** the delay between LED pulse and polymer laser pulse in different pump intensity regions; **c** and **d** dynamics of output pulses from LED (*black*) and polymer laser (*red*) with the pump intensity in region 2 and 3, respectively

5.8.3 Laser Beam Divergence Study

The BBEHP-PPV laser output beam divergence was measured with an LED peak intensity of 840 W/cm^2. The beam divergence was found to be approximately 5 mrad in the direction perpendicular to the axis of the fan-shaped output, by measuring the polymer laser beam profile at various distances using a CCD beam profiler (Fig. 5.18).

An angle-dependent spectrum measurement was set-up in the way that described in Chap. 3. LED pumped BBEHP-PPV DFB laser emission spectra were collected with a fibre (aperture less than 0.5 mm) in a range of ±5°, while the LED peak intensity was far above lasing threshold. The distance between the polymer laser and the fibre end was approximately 7 cm when the fibre was positioned at 0° to the laser surface. Figure 5.19 shows the angle-resolved emission from the LED pumped polymer laser. In Fig. 5.19 (a), A and B represent the light in TE_0 mode subsequently Bragg scattered out of the polymer waveguide by the grating, while position C represents a stop-band at 535 nm. The high-intensity lasing peak appears at the long-wavelength side (position D). In Fig. 5.19 (b), the laser intensity profile is used to estimate the beam divergence, which is approximately 5 mrad.

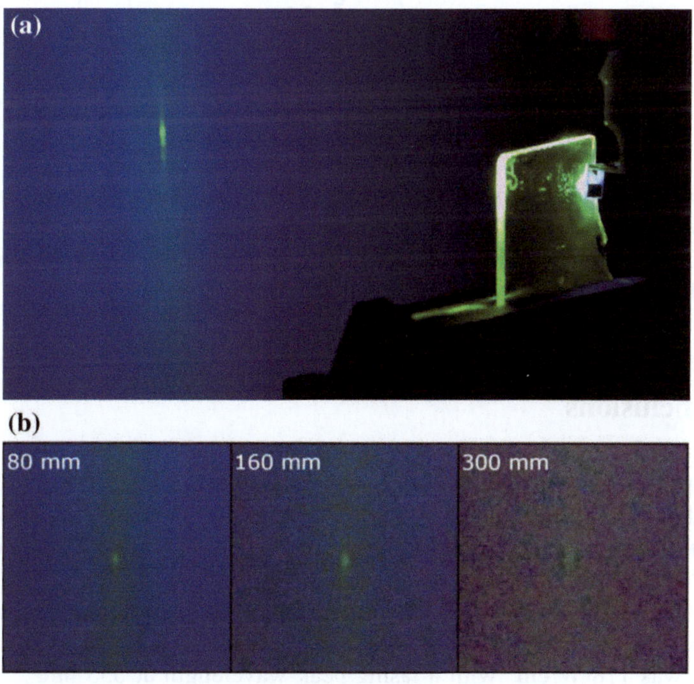

Fig. 5.18 **a** Photo of an LED pumped BBEHP-PPV laser with LED power 1.7-time of the laser threshold; **b** laser emission patterns at three different distances away from the laser, (*left*) 80 mm, (*middle*) 160 mm and (*right*) 300 mm

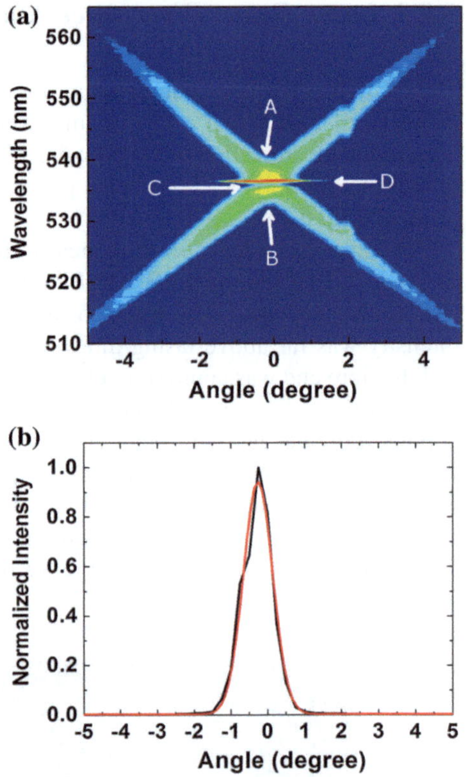

Fig. 5.19 **a** Angle-dependent emission map; *red-coloured* regions represent the strong emission and *blue-coloured* regions represent the weak emission. Labels A to D refer to: A, the long-wavelength emission band; B, the short wavelength emission band; C, the dip in emission at the photonic band gap and D, the narrow-linewidth lasing; **b** angle-resolved intensity profile (*black*) at the lasing wavelength of 536.7 nm with a Gaussian fitting (*red*)

The radiation pattern of the commercial LED was very close to Lambertian under the pulse operation. Therefore, with the LED pumped polymer laser, we successfully converted an incoherent and divergent light source into a coherent and highly directional laser source.

5.9 Conclusions

Eight batches of BBEHP-PPV polymer have been studied for making low threshold lasers in this chapter. Measurements of PLQY and ASE show that the variation in the molecular weight and polydispersity can affect the optical properties of the polymer (Table 5.6).

Very low threshold BBEHP-PPV lasers were fabricated from the best batch with one-dimensional DFB resonators. The lowest threshold achieved with OPO pumping was 118 W/cm^2 with a lasing peak-wavelength at 533 nm.

Distributed feedback polymer lasers pumped by commercial InGaN light-emitting diode have been demonstrated. The LED peak intensity driven under 10 s

Table 5.6 Summary of the film PLQY and molecular weight of all the BBEHP-PPV batches with their ASE thresholds

Batch name	Film PLQY (%)	Mw	ASE threshold (W/cm^2)
AK342P	82	582,800	300
AK350DCM	77	455,300	270
AK350Me2CO	82	28,430	490
AK391DCM2	81	205,300	220
AK391DCMB	84	341,100	170
AK391DCMF	84	595,100	640
AK408DCM	86	543,400	280
AK408R	88	472,800	270

of nanosecond pulses were characterized. At 20 Hz repetition rate, the blue-emitting LED could provided up to 1,000 W/cm^2. For a LED pumped BBEHP-PPV laser, the narrow-linewidth polymer laser peak appeared at 533 nm for peak drive currents above 15 A. The output beam and pulse dynamics of the laser were characterized, along with a 'double-threshold' phenomenon in such long-pulse pumping geometry, attributed to the slow turn-on of the LED pulse.

This is the first time that a polymer laser has been pumped by a commercial LED far above threshold to show a clear laser beam. The LED pumped BBEHP-PPV laser provides a compact and versatile laser device with green emission. Such a demonstration is encouraging for the prospects of using plastic lasers in emerging practical applications, such as trace explosive sensing [14], with substantial reduction in the complexity, size and expense of the optical sources. Furthermore, the pulse duration of the LEDs can span a wide range, from tens of nanoseconds to CW operation, entirely depending on its driving conditions. Thus the LED pumped organic laser configuration opens a convenient way to study polymer laser behaviour under long pulse pumping. More important in the HYPIX project, the success of using commercial LEDs as a pump source provides essential insight for the subsequent integration of micro-LEDs and polymer lasers that will be discussed in the next chapter.

References

1. Samuel, I. D. W., & Turnbull, G. A. (2007). Organic semiconductor lasers. *Chemical Reviews, 107*(4), 1272–1295.
2. McGehee, M. D., & Heeger, A. J. (2000). Semiconducting (conjugated) polymers as materials for solid-state lasers. *Advanced Materials, 12*(22), 1655–1668.
3. Chénais, S., & Forget, S. (2011). Recent advances in solid-state organic lasers. *Polymer International, 61*(3), 390–406.
4. Yang, Y., Turnbull, G. A., & Samuel, I. D. W. (2008). Hybrid optoelectronics: A polymer laser pumped by a nitride light-emitting diode. *Applied Physics Letters, 92*(16), 163306.
5. Flory, P. J. (1941). Molecular size distribution in three dimensional polymers I. Gelation. *Journal of the American Chemical Society, 63*(11), 3083–3090.

6. Clarke-Pringle, T. L., & MacGregor, J. F. (1998). Optimization of molecular-weight distribution using batch-to-batch adjustments. *Industrial and Engineering Chemistry Research, 37*(9), 3660–3669.
7. Donley, C. L., Zaumseil, J., Andreasen, J. W., Nielsen, M. M., Sirringhaus, H., Friend, R. H., et al. (2005). Effects of packing structure on the optoelectronic and charge transport properties in poly(9,9-di-n-octylfluorene-alt-benzothiadiazole). *Journal of the American Chemical Society, 127*(37), 12890–12899.
8. Koynov, K., Bahtiar, A., Ahn, T., Cordeiro, R. M., Horhold, H. H., & Bubeck, C. (2006). Molecular weight dependence of chain orientation and optical constants of thin films of the conjugated polymer MEH-PPV. *Macromolecules, 39*(25), 8692–8698.
9. Winfield, J. M., Donley, C. L., & Kim, J. S. (2007). Anisotropic optical constants of electroluminescent conjugated polymer thin films determined by variable-angle spectroscopic ellipsometry. *Journal of Applied Physics, 102*(6), 063505.
10. Meier, M., Mekis, A., Dodabalapur, A., Timko, A., Slusher, R. E., Joannopoulos, J. D., et al. (1999). Laser action from two-dimensional distributed feedback in photonic crystals. *Applied Physics Letters, 74*(1), 7–9.
11. Barlow, G. F., Shore, A., Turnbull, G. A., & Samuel, I. D. W. (2004). Design and analysis of a low-threshold polymer circular-grating distributed-feedback laser. *Journal of the Optical Society of America B, 21*(12), 2142–2150.
12. Turnbull, G. A., Andrew, P., Barnes, W. L., & Samuel, I. D. W. (2003). Operating characteristics of a semiconducting polymer laser pumped by a microchip laser. *Applied Physics Letters, 82*(3), 313–315.
13. Samuel, I. D. W., Namdas, E. B., & Turnbull, G. A. (2009). How to recognize lasing. *Nature Photonics, 3*(10), 546–549.
14. Rose, A., Zhu, Z. G., Madigan, C. F., Swager, T. M., & Bulovic, V. (2005). Sensitivity gains in chemosensing by lasing action in organic polymers. *Nature, 434*(7035), 876–879.

Chapter 6
Low Threshold Nanoimprinted Organic Lasers Integration with Micro-LED Arrays

6.1 Introduction

To date there has been no demonstration of electrically driven lasing in the organic semiconductors due to problems including charge-induced absorption and metal contact loss, discussed in Chap. 2. Therefore, the development of integrated organic optoelectronic devices, with high-performance, for potential practical applications including sensing and lighting is challenging. Indirect electrical pumping is one solution to this challenge as described in Chap. 5 where organic polymer lasers have been successfully pumped far above threshold by a pulsed commercial InGaN LED.

In the HYPIX project, the vision is to fully hybridise organic semiconductor optoelectronics with inorganic complementary metal-oxide semiconductor (CMOS) based electronic control circuitry, by using an interface of gallium nitride (GaN) optoelectronics, i.e. micro-pixellated GaN LEDs. To this end, the prospect of using micro-LEDs to pump the polymer laser is studied in this chapter. In Sect. 6.2 the early generations of micro-LED designs are presented and their output power intensities are characterised. In Sect. 6.3 the third generation micro-LED arrays with a new stripe geometry and the improvement in power capability are discussed. In Sect. 6.4 the polymer laser thresholds with different pump geometries are investigated using OPO as the excitation source. The initial micro-LED stripe-array pumped organic laser is studied in Sect. 6.5.

As introduced in Chap. 3, nanoimprint lithography provides an increased fabrication through-put (i.e. a parallel process unlike electrical beam lithography which is a serial process) and a high resolution technique to make micro-resonators for organic semiconductor lasers, with the potential for roll-to-roll fabrication. To optimize the polymer laser threshold, mixed-order gratings for BBEHP-PPV lasers are fabricated with UV-nanoimprint lithography, described in Sect. 6.6. Ultra-low thresholds of the nanoimprinted mixed-order lasers are demonstrated in Sect. 6.7, and Sect. 6.8 presents the performance of these low threshold polymer lasers when pumped by micro-LED arrays, followed by conclusions in Sect. 6.9.

Y. Wang, *Low Threshold Organic Semiconductor Lasers*, Springer Theses,
DOI: 10.1007/978-3-319-01267-4_6,

6.2 Pulsed Micro-LED Power Characterization

All the micro-LED devices used in this work were fabricated at the Institute of Photonics (IoP), University of Strathclyde in Glasgow. The intensity of the devices under pulsed operation were measured at St Andrews before being used for laser pumping. Dr. Zheng Gong, Dr. Jonathan Mckendry, Dr. David Massoubre and Mr. Johannes Herrnsdorf at the IoP have put great efforts into the micro-LED fabrication processes; Dr Erdan Gu and Prof. Martin Dawson were also closely involved in the discussions throughout the development of the micro-LED devices.

The IoP group use gallium nitride (GaN) in micro-LED fabrication because of the ultraviolet to blue light output of GaN, which is compatible with the absorption characteristics of many organic semiconductor gain media. The micro-LEDs were fabricated from commercially available wafers grown in patterned sapphire substrate, which were available at peak emission wavelengths of approximately 370, 405 and 450 nm. The cross section of the micro-LED structure is illustrated in Fig. 6.1a. The external quantum efficiency and maximum output power under CW operation mode increased as the wavelength increased to 450 nm. Therefore, in this work we chose the 450 nm devices, which also matched well with the absorption of the low threshold polymer BBEHP-PPV used in Chap. 5. The typical emission spectrum of the 450 nm micro-LED had a linewidth of 20 nm (FWHM) and overlapped extremely well with the BBEHP-PPV's absorption spectrum, as shown in Fig. 6.1b.

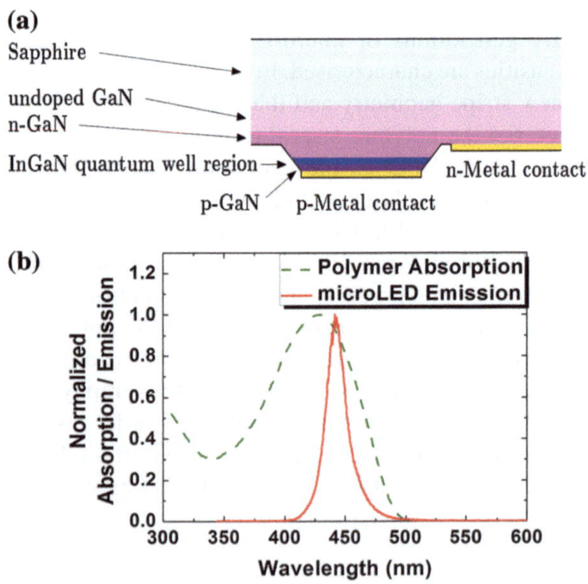

Fig. 6.1 **a** Cross section of a flip-chip micro-LED structure with InGaN quantum wells; adapted from Ref. [3]; **b** normalized absorption spectrum of BBEHP-PPV (AK408DCM) and emission spectrum of the 450 nm micro-LED

6.2.1 First Generation Device

The first generation micro-LED device consisted of 30 μm diameter circular pixels on a 35 μm pitch (fill factor of 58 %). For a single pixel, the maximum pulse energy could reach 500 pJ for the 450 nm devices. With a pixel area of 7.1×10^{-6} cm^2, the intensity from one pixel was approximately 400 W/cm^2. This value is comparable with most of the commercial LEDs' power capabilities; however, the commercial LED has an emitting area of about 10^{-2} cm^2 and so far we have not considered the divergence of the single-pixel micro-LED. The intensity at the surface of the sapphire substrate (350 μm) was estimated to be reduced to 2.5 W/cm^2. For the purpose of using it as a laser-pumping source, a single pixel of the micro-LED was not practical as its low power density and miniature emitting area, which could not supply a sufficient gain volume. Previous work in St Andrews on a copolymer laser using ADS233YE found that the DFB laser threshold increased dramatically when the excitation spot was smaller than 1 mm in diameter [1].

For pumping polymer lasers, the micro-pixels of the LEDs were interconnected to form arrays of micro-LEDs, which could then be synchronously pulsed. On a sapphire substrate there were different sizes of interconnected square and rectangular arrays in the first generation devices, as shown in Fig. 6.2. They had independent p-contacts but a shared n-contact. The largest array in the device was a 15-by-15 pixel array, forming a 520 by 520 μm square (2.7×10^{-3} cm^2).

The output power intensities from these 520 by 520 μm arrays, driven by the laser diode driver (PCO-7110-120-15) at repetition rate of 20 Hz, were characterized. The maximum power from the large array could reach 70 W/cm^2 with a 21 ns (FWHM) pulse before any damage occurred to the micro-LEDs. The damage happened either to the individual pixels or the bonding wires, due to the very high pulsed currents flowing through the device (a few kA/cm^2). Usually a

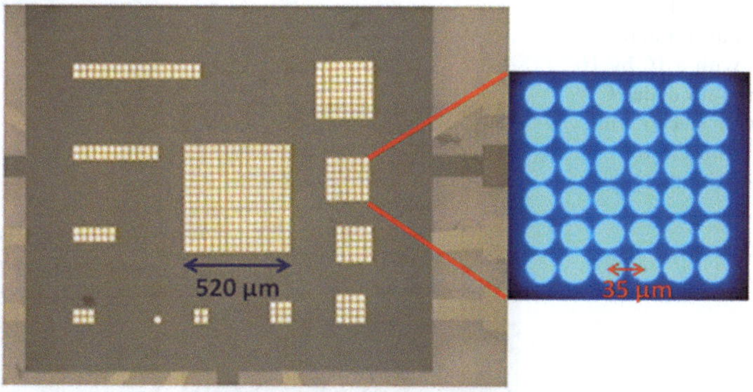

Fig. 6.2 Micrograph of top view of the sapphire substrate of the interconnected micro-LED arrays with a 6-by-6 array emitting under CW operation; the micro-LED chip was glued onto the backside of a piece of printed circuit board (PCB) and then wire-bonded to the metal pads on the PCB; reproduced with permission from the IoP, University of Strathclyde

sudden drop of the LED power was a sign of irreversible damage to the device. One of the reasons for the lower output intensity of the large area array compared to the single pixel (70 W/cm^2 compared to 400 W/cm^2) was the poor heat dispersion in the device under the high-current pulse-operation [2]. The low fill factor in these arrays was another reason for low total intensity.

The intensity from the first generation micro-LED arrays was not sufficient for pumping the BBEHP-PPV laser above threshold. Several approaches were then developed to improve the micro-LED power capacity hence the second generation.

6.2.2 Second Generation Device

The main change in the second generation design of the micro-LEDs was an increase in fill factor. In Fig. 6.3, we can see that the pixels are square rather than circular increasing the fill factor from 58 to 83 %. The pixel size was 30 by 30 μm with a 33 μm pitch.

The 16 by 16 array had a maximum output intensity of 225 W/cm^2 (see Fig. 6.4) with an illuminated area on the sapphire surface of 2.6×10^{-3} cm^2. For a similar size array, the intensity improved three times due to the much higher filling factor in the second generation devices.

To improve the thermal characteristics of the micro-LED chips, some devices were bonded to a diamond-backplane as diamond has an improved heat-spreading ability compared to PCB. Alternatively, some wire-bonded micro-LED chips were attached to a ceramic heat sink by electrically-insulating thermal grease, at the IoP, as shown in Fig. 6.5. Under CW operation, performances of both types of devices in comparison with the same-size array without any additional heat-spreading elements were investigated by the IoP group [3]. From Fig. 6.6 we can see that

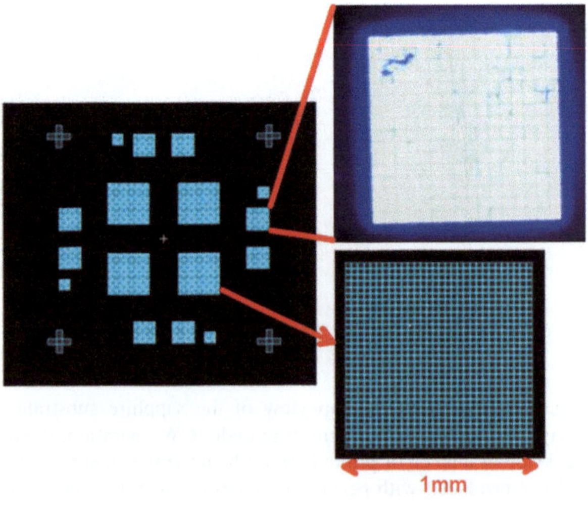

Fig. 6.3 Design of the second generation micro-LED arrays with a 16 by 16 array, lighting up under CW operation; reproduced with permission from the IoP, University of Strathclyde

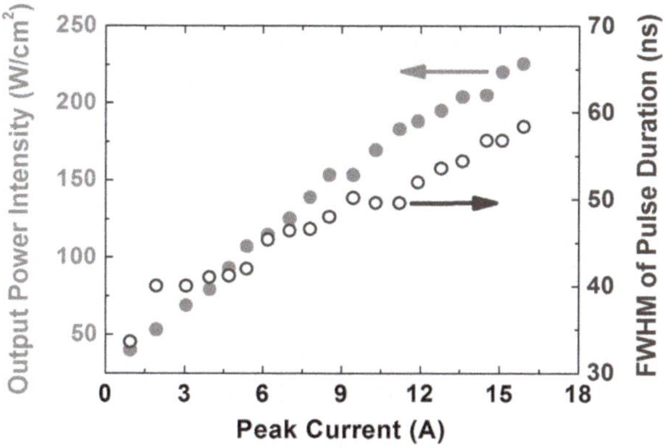

Fig. 6.4 Intensity output from a 16-by-16 pixel micro-LED array and the optical pulse duration as a function of driving current

Fig. 6.5 Schematic (*top*) and photographic (*bottom*) images of a micro-LED array with heat sink; reproduced with permission from the IoP, University of Strathclyde

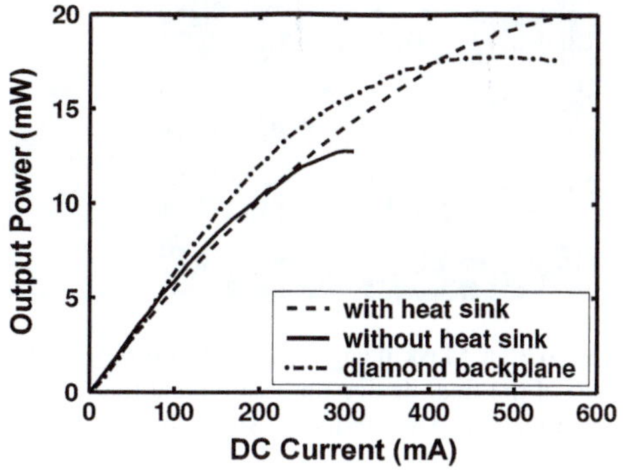

Fig. 6.6 DC power characterization with (*dashed line*) and without (*solid line*) heat sink, compared to the device with diamond backplane (*dashed* and *dotted line*); data obtained by Johannes Herrnsdorf at the IoP [3]

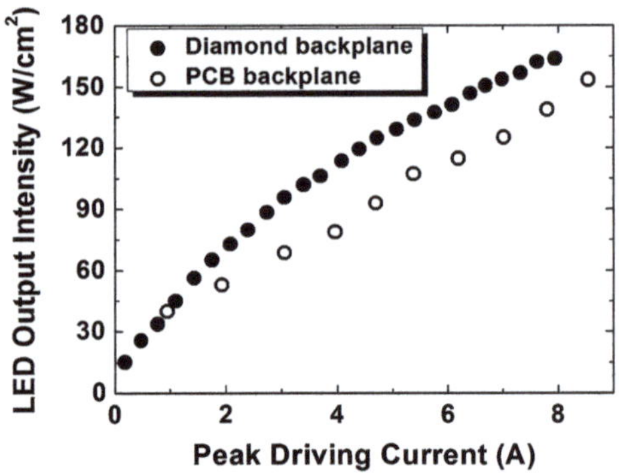

Fig. 6.7 Comparison of pulsed output intensity between micro-LED device with (*closed circle*) and without (*open circle*) diamond backplane

both the heat sink and diamond backplane extend the damage current of these micro-LED devices and improve the maximum DC output power. However, under pulse operation the improvement to the output intensity (characterized at St Andrews) with the diamond backplane was not obvious, see Fig. 6.7.

6.3 Third Generation Micro-LEDs

6.3.1 New Geometry Design

Using the custom micro-pixelated GaN LEDs, different geometries are more achievable compared to commercial LEDs. Thin and long stripes, in principle, could improve one-dimension heat dispersal in the devices. Therefore, the design of the third generation of the micro-LED devices was focused on stripe geometries, see Fig. 6.8.

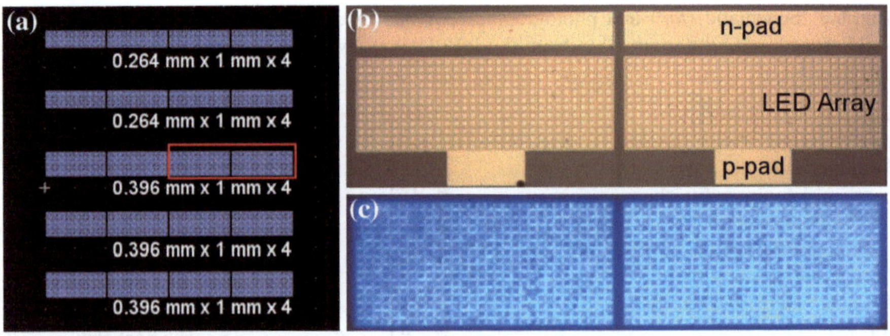

Fig. 6.8 **a** Mask designed for the third generation micro-LED device; **b** and **c** photographs of part of the final device that equivalent to the highlighted region in **a**; reproduced with permission from the IoP, University of Strathclyde

The pixel size and fill factor were kept the same as the second generation, i.e. 30 by 30 μm with a 33 μm pitch. The micro-LEDs were interconnected to form isolated units, four per stripe as seen in Fig. 6.8a, which were either an 8 by 30 array or a 12 by 30 array. This means that the unit lengths are all 1 mm, but the array widths are different. These stripe-shaped micro-LED arrays were designed in such a way that they could be either driven by a single driver or by two independent drivers. In other words, the arrays had independent n-contact/p-contact. In this way we could choose the stripe length to be 1, 2, 3 or 4 mm, for example in Fig. 6.8c, a 2 × 1 mm × 0.396 mm stripe was operated under CW mode.

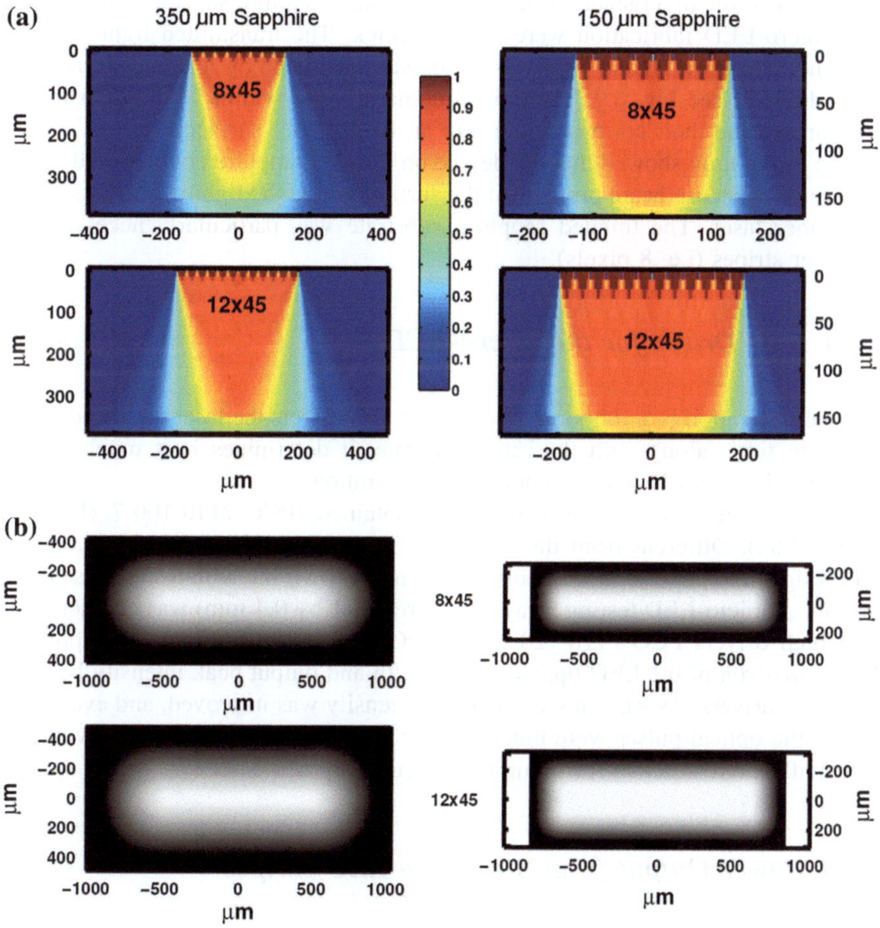

Fig. 6.9 **a** Calculated intensity of light as it diverges while propagating through the sapphire window from the micro-LED stripe arrays using a Monte-Carlo ray-tracing model; **b** calculated intensity profiles at the sapphire surface; data obtained by Johannes Herrnsdorf at the IoP [3]

Table 6.1 Summary of the stripe width at sapphire surface

Sapphire thickness (μm)	Stripe pixel number and width	Width at sapphire surface (μm)
350	$8 \times n$ (264 μm)	365
350	$12 \times n$ (396 μm)	420
150	$8 \times n$ (264 μm)	267
150	$12 \times n$ (396 μm)	400

6.3.2 Thinned Sapphire Substrate

Another step to maximize the output intensity of the micro-LED arrays accessible to the organic laser was to thin down the sapphire substrate to 150 μm to decrease the impact of the divergence of the LED. Standard sapphire substrates used in the GaN micro-LED fabrication were 350 μm thick. The transmitted light intensity from the stripes with 8 and 12 pixels width was calculated by Johannes Herrnsdorf at the IoP [3], see Fig. 6.9. Table 6.1 summarizes the widths of different micro-LED arrays and their corresponding size at the sapphire surface.

The calculation showed that the device on thinned sapphire could provide more uniform and intense beam profiles on the surface where the light was accessible by a polymer laser. The thinned sapphire substrate was particularly necessary for narrower stripes (i.e. 8 pixels).

6.3.3 New Driver for the Micro-LEDs

The pulse duration is an important parameter as it affects the laser threshold and at the same time, along with the repetition rate, it determines how much heat is generated in the LED devices under pulse operation.

For these reasons, a new driver was obtained, PCO-7110-100-7 (Directed Energy Inc.). Different from the driver PCO-7110-120-15 used before, this one was designed to provide shorter pulses ($\sim$7 ns FWHM) but with lower pulse peak-current. A micro-LED testing array (total area of 1 by 0.4 mm) was characterized with both drivers PCO-7110-120-15 and PCO-7110-100-7. Figure 6.10 presents the comparison of the LED optical pulse width and output peak intensity between these two drivers. As we can see, the peak intensity was improved, and even more clearly the optical pulses were much shorter with the new driver, which would be potentially more suitable for pumping organic polymers lasers.

6.3.4 Power Output from the Micro-LED Stripes

Table 6.2 summarizes some of the best performing 450 nm stripe-arrays with different length and width driven by the PCO-7110-100-7 driver at 20 Hz. It can be seen that the smaller arrays (shorter or thinner) produced higher intensity.

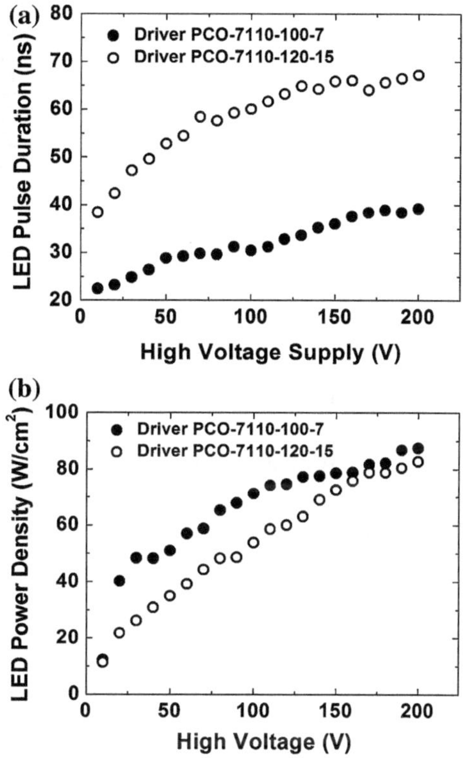

Fig. 6.10 Comparison between two drivers with 1 mm micro LED stripe (device LS-450-7 p21 n2)

Table 6.2 Summary of the output power performances of the stripe micro-LED arrays

Device	Dimension (mm^2)	Area on Sapphire (cm^2)	Maximum peak intensity[a] (W/cm^2)
LS-450-8	$2 \times (1 \times 0.27)$	5.4×10^{-3}	189
LS-450-10	1×0.40	4.0×10^{-3}	232
LS-450-10	1×0.27	2.7×10^{-3}	290
LS-450-10	3 x (1×0.40)	1.2×10^{-2}	146
LS-450-10	3 x (1×0.27)	8.1×10^{-3}	166
LS-450-11	2 x (1×0.40)	8.0×10^{-3}	157

[a] The actual maximum peak intensity could be increased approximately by a further 20 % just before the damage happens to the micro-LEDs

Compared to a similar size square array from the second generation device in Sect. 6.2.2 (16 by 16 pixels, 2.6×10^{-3} cm^2) which gave an intensity of 225 W/cm^2, the 30-by-8-pixel array (1 mm by 0.27 mm, 2.7×10^{-3} cm^2) produced 1.3 times higher peak intensity, of 290 W/cm^2.

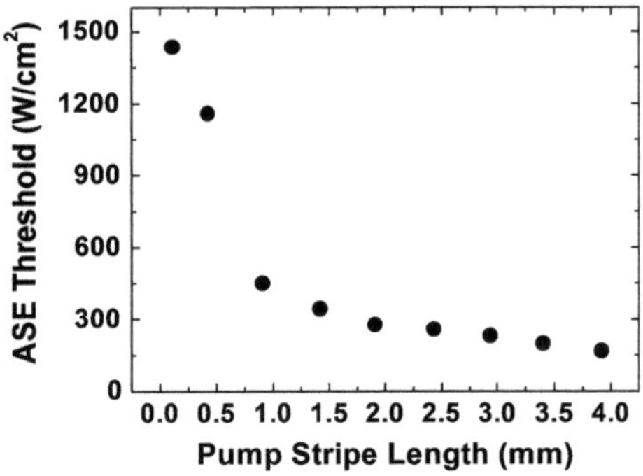

Fig. 6.11 ASE threshold as a function of pump stripe length

6.4 Pump Beam Shape Dependent Lasing Threshold

The BBEHP-PPV ASE and laser thresholds using stripe-shaped pump beams were investigated with 4 ns FWHM 450 nm pulses from the OPO at 20 Hz repetition rate. A combination of a cylindrical lens and an adjustable slit were used to vary the dimension of the excitation beam.

In the ASE measurement, the pump beam was shaped into a 440 μm wide stripe, and the length of the stripe was varied between 0 and 3.9 mm. As we can see from Fig. 6.11, the ASE threshold of the BBEHP-PPV film increased as the pump stripe was shortened. The threshold increased dramatically when the pump stripe was shorter than 1 mm.

Surface emitting second-order DFB lasers were made with the 355 ± 5 nm period holographic silica grating used in Chap. 5. Figure 6.12 shows the BBEHP-PPV laser threshold decreases by more than an order of magnitude as the pump stripe length was changed from 0.5 to 2.5 mm, with the width of the stripe fixed at 420 μm. With the stripe length in the range of 2.5–5 mm, the trend of the decreasing threshold flattens. The lowest lasing threshold was found to be 45 W/cm^2 with a 3.9 mm by 420 μm pump stripe. When the stripe length was further increased to 7.8 mm, the laser threshold appeared to increase to just above 100 W/cm^2, which could be due to a defect in the holographic gratings over such a long distance/area.

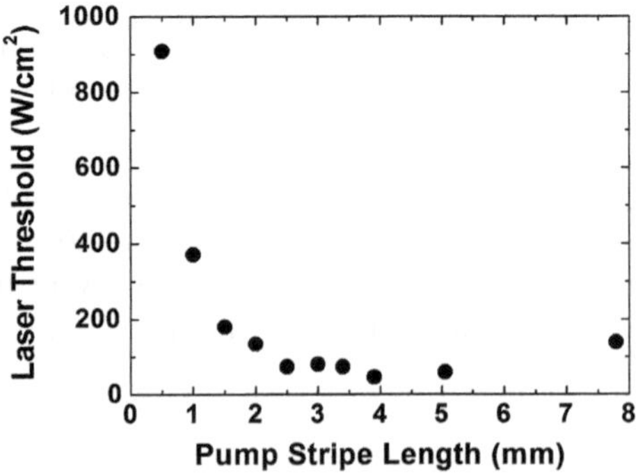

Fig. 6.12 Laser threshold as a function of pump stripe length

6.5 Towards Micro-LED Pumped BBEHP-PPV Laser

With the laser threshold improved in the stripe pumping experiments, it was promising that the low threshold DFB laser could be pumped with the long stripe arrays in the third generation micro-LED devices.

A BBEHP-PPV laser was made with the holographic silica grating and was mounted in front of a 2 × 1 mm × 0.27 mm micro-LED stripe, pulsed by the PCO-7110-100-7 driver at 20 Hz. The micro-LED peak intensity and pulse width as a function of the driving current are shown in Fig. 6.13a. At low pump intensities, a Bragg dip was evident at around 530 nm. As the LED power increased, a sharp peak began to rise at 531 nm, on the right-side of the stop-band, see Fig. 6.13b. The relative emission intensity at this wavelength was plotted against the input peak intensity in Fig. 6.13c. The slope efficiency started to change when the LED intensity was above 170 W/cm^2. Both the spectra and the slope change suggested the on-set of lasing. However, the micro-LED could not provide more power.

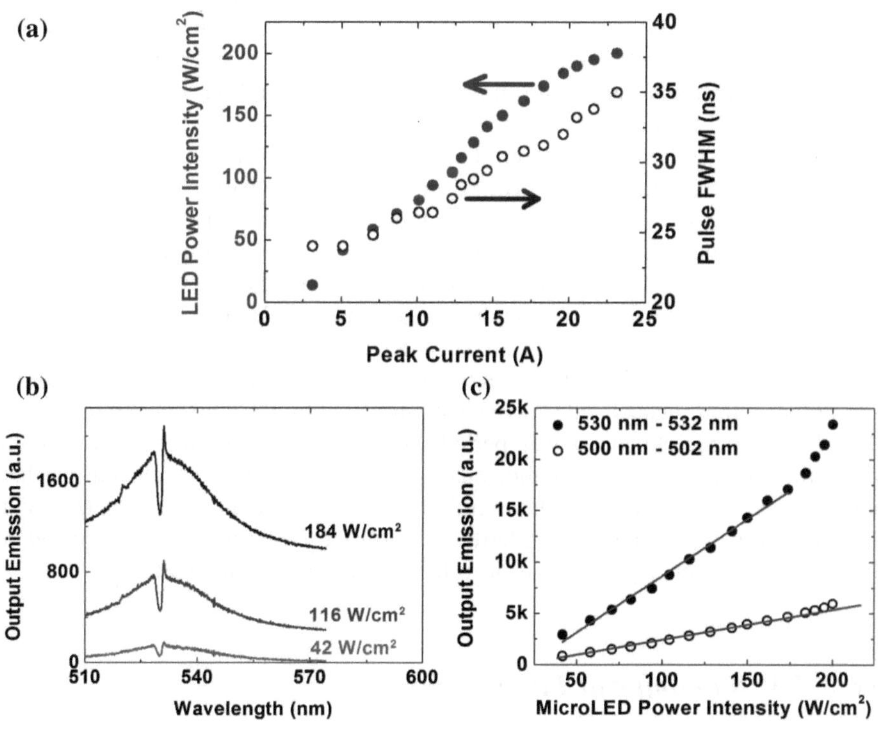

Fig. 6.13 **a** Output intensity (closed circles) from a 2 × 8 × 30-pixel micro-LED array and the optical pulse duration (open circles) as a function of driving current; **b** The micro-LED pumped polymer laser emission spectra at three different power intensities of the micro-LED array, 42, 116 and 184 W/cm^2; **c** output intensity at the wavelength of 531 and 501 nm (away from the Bragg dip) as a function of micro-LED peak intensity

6.6 Laser Resonator Development

The output intensity from the micro-LED devices has improved dramatically throughout the HYPIX project, from 70 W/cm^2 to over 250 W/cm^2. To achieve the goal of a micro-LED pumped polymer laser, the polymer lasing threshold have also been optimised by developing the polymer laser resonators.

6.6.1 Mixed-Order DFB Resonator

As mentioned in Chap. 2, the efficient surface emission from a second-order DFB laser represents a loss mechanism in the laser system, while for a first-order DFB resonator strong feedback with little output consequently permits low threshold operation. However, the edge emission from the first-order resonator has its own limits including poor beam quality. The idea of a mixed-order resonator [4] takes the advantages of the two approaches, i.e. controllable vertical outcoupling of the laser radiation from the second-order region and strong feedback from the first-order region, potentially reducing the threshold. In Karnutsch et al's work, the width of the second-order region was varied from 4 to 178 grating periods for threshold comparison, and with more than 30 outcoupling periods the laser threshold started to increase significantly.

As in the conventional DFB laser, the optical feedback wavelength of the mixed-order laser is determined by the Bragg equation. In this work, with the polymer BBEHP-PPV as the gain medium, the appropriate period for outcoupling the preferable wavelength (around 535 nm) is 350 nm and for in-plane feedback is 175 nm. Figure 6.14 shows a schematic drawing of a mixed-order grating with four periods in the second-order outcoupling region. Using electron beam lithography followed by reactive-ion etching, mixed-order gratings of dimension 1 by 1 mm, with 4, 8, 15 and 30 periods in the second-order region respectively and equal-size first-order regions on each side, have been fabricated by Kelvin Nanotechnology in Glasgow. In principle, the fewer periods of second-order grating the lower the lasing threshold, due to the smaller outcoupling loss. On the other hand, the less output coupling also means the laser emission is more difficult to be detected. With the mixed-order gratings, a standard second-order DFB grating with period of 350 nm as a reference grating was also fabricated on the same wafer.

6.6.2 UV Nanoimprint Lithography

The original high-quality resonator structures were fabricated in a silicon wafer. UV nanoimprint lithography (UV-NIL) [5, 6] was used to replicate the master structures. A monolayer of anti-sticking material, perfluorooctyl-trichlorosilane

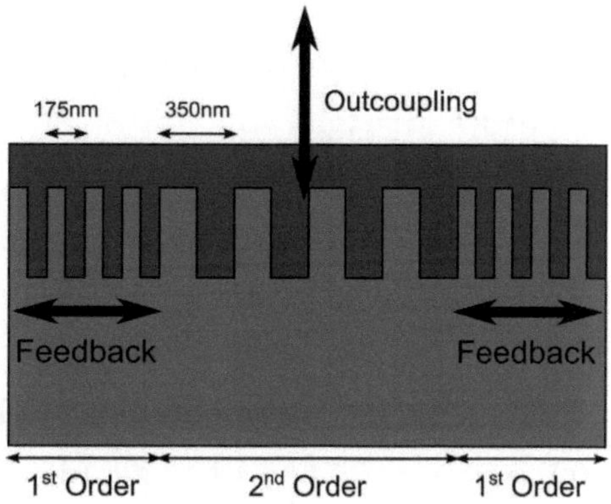

Fig. 6.14 Schematic drawing of mixed-order gratings with four periods of second-order gratings in the outcoupling region surrounded by first-order gratings on both sides

(Sigma Aldrich), was used to coat the master wafer first, as described in Chap. 3. A soft stamp was made on a glass substrate as a replica of the silicon master and then this was used to transfer the nanostructures into a UV-NIL resist mr-UVCur06 (Microresist Technologies) using the EVG620 photomask aligner with custom tooling for UV-NIL. Glass substrates for UV-NIL were cleaned within a piranha solution, 3:1 mixture of concentrated sulfuric acid (H_2SO_4) with hydrogen peroxide (H_2O_2) for 5 min. The piranha etching would remove the organic residue on the glass and more importantly make the substrate hydrophilic; this was to ensure a thin and uniform adhesion promoter layer mr-APS1 (around 10 nm, post-bake for 60 s at 150 °C) to be deposited on to the glass substrate increasing the adhesion of the UV-NIL resist to the glass. Then an approximately 200 nm thick mr-UVCur06 resist layer was spin-coated at a speed of 3,500 rpm for 60 s, post-baked at 100 °C for 30–40 s, on the top of mr-APS1. The soft stamp was then placed in contact with the UV-NIL resist-coated glass substrate by a set vacuum of 950 mbar. The imprint pressure and time were critical during the transfer of the nano-scaled patterns. At the same time, the UV resist was cured using UV illumination (320–420 nm) for 220 s at a dose of approximately 1 J/cm^2. After UV exposure, the soft stamp was detached and the structure was finally transferred into the resist layer.

Prior to completing the fabrication of the polymer laser, the final imprinted structures in mr-UVCur06 were characterized using scanning electron microscopy (SEM) and atomic force microscopy (AFM). Figure 6.15 shows a comparison of topography images for the master and imprint structures. From Fig. 6.15a and b, we can see that the silicon master grating, produced with the e-beam lithography, is of good quality, with 350 nm period for the outcoupling region and 175 nm period for the feedback region. The anti-sticking monolayer does not affect the topographic profile of the nanostructures. From Fig. 6.15c and d, we can see that the master gratings have roughly 50 % duty cycle (ratio of ridge-width and period) with 105 ± 5 nm groove-depth for second-order gratings. The first-order gratings are shallower (65 ± 5 nm), possibly due to the slower etching rate for these sub-hundred-nm grooves during the master fabrication process. The gentle curving of

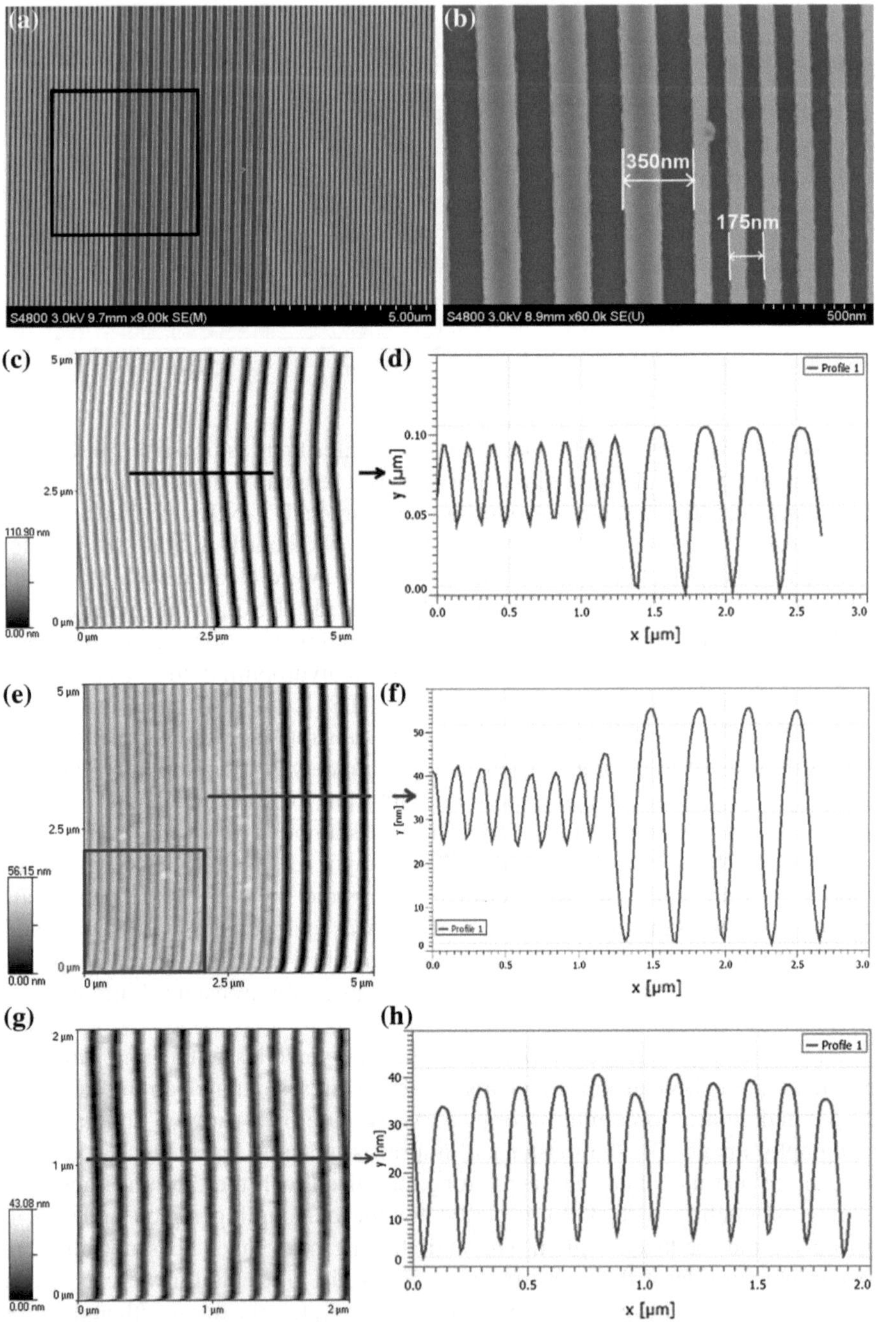
(a)
S4800 3.0kV 9.7mm x9.00k SE(M)
5.00um
(b)
350nm
175nm
S4800 3.0kV 8.9mm x60.0k SE(U)
500nm
(c)
(d)
Profile 1
y [µm]
x [µm]
(e)
(f)
y [nm]
x [µm]
(g)
(h)
y [nm]
x [µm]

◀ **Fig. 6.15** **a** SEM images of the master structure with 15 periods of second order gratings sandwiched by first-order gratings; **b** SEM images of the master gratings after anti-sticking monolayer treatment showing the periods of the gratings; **c** Contact mode AFM image of the highlighted 5 by 5 μm area in **a**, with **d** the line profile extracted showing the groove-depths; **e** AFM image of imprinted gratings in mr-UVCur06 in a 5 by 5 μm area, with **f** the line profile extracted; **g** AFM image of a 2 by 2 μm area of the first order gratings highlighted in **e**, with **h** the line profile showing the groove depth of the first-order gratings is about 35 ± 5 nm

the gratings in the AFM images was due to the drift of the probe tip/cantilever. It is worth mentioning that the depth of the gratings may well be underestimated due to the limit imposed by the shape of the AFM tip, i.e. the resultant AFM image consists of a convolution of both the real grating shape and the AFM tip shape. To characterize the grating depth more precisely, it is suggested to perform the imprint process on a silicon substrate and then cleave the silicon gratings in order to measure the grating profile, i.e. actual depth using the SEM. Figure 6.15e–h show that the final structures imprinted in the UV resist, replicating the period, duty cycle and uniformity of the master structures in silicon very well, although the groove depth reduces to 55 ± 5 nm for second-order gratings and 35 ± 5 nm for the first-order gratings. A higher resolution scan was performed in a smaller area of the first order grating region to confirm the depth of the sub-hundred-nm groove, see Fig. 6.15g.

To finish the laser fabrication, a BBEHP-PPV polymer film was deposited by spin-coating on top of the patterned UV-resist layer, and finally capped with a ~1.5 μm thick CYTOP layer. The refractive indices of the adhesion promoter mr-APS1 and the resist mr-UVCur06 were measured using ellipsometry. For the wavelength of 535 nm, n_{adh} is 1.48 while n_{resist} is 1.50. With a 250 nm thick BBEHP-PPV film as the laser gain medium, the TE_0 mode intensity profile is plotted in Fig. 6.16b. The effective refractive index of the TE_0 mode at 535 nm was calculated to be 1.53.

6.7 Ultra-Low Threshold Nanoimprinted Laser

The UV-NIL mixed-order BBEHP-PPV (AK408R) lasers were first optically excited by the OPO at 450 nm. The lasers were characterized with a circular excitation spot and a stripe pump beam, respectively. A UV-NIL laser with 30 periods of second-order gratings had a threshold of 62 W/cm^2 with a 1.1 mm diameter circular pump beam and the same laser pumped with 1 by 0.24 mm stripe beam had a similar lasing threshold of 58 W/cm^2, see Fig. 6.17a and b, indicating that the pump beam geometry did not affect the threshold in these mixed-order DFB lasers.

On the same sample and pumped with the same-size stripe, the second-order reference laser had a threshold of 95 W/cm^2. This suggested that the laser made

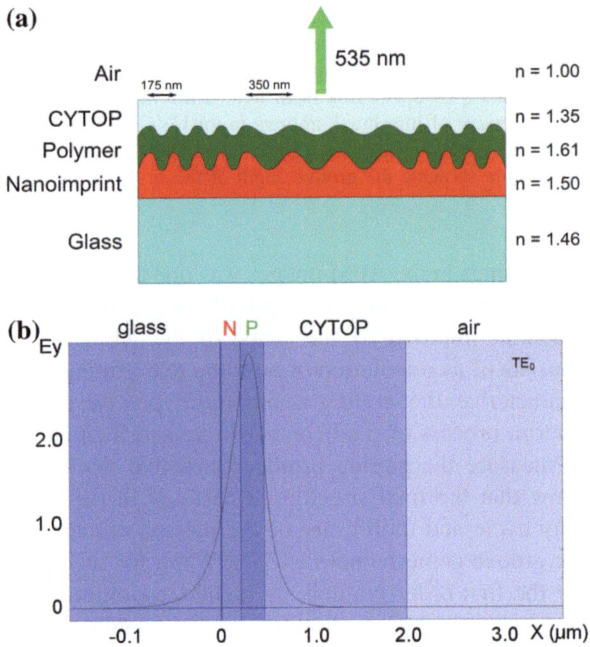

Fig. 6.16 **a** Schematic drawing of a UV-nanoimprinted mixed-order BBEHP-PPV polymer laser encapsulated using CYTOP, with the refractive index value of each layer; **b** intensity profile of the fundamental TE mode in the multi-layer laser system, plotted using an online source: http://wwwhome.math.utwente.nl/~hammerm/oms.html; **N** stands for the nanoimprint resist with the adhesion promoter and P stands for the BBEHP-PPV polymer

with the unconventional mixed-order DFB resonator had an approximately 40 % reduction in threshold compared to the laser made with the standard second-order resonator.

Different from the second-order laser, Fig. 6.17d shows a three-mode lasing spectrum when the mixed-order laser is pumped above threshold. Each peak had a linewidth of ~0.1 nm. The peaks at the shortest (534.1 nm) and longest (537.1 nm) wavelength were the two Bragg-scattered band-edge modes that originated from the periodic nature of the gratings, commonly seen with first-order and second-order DFB lasers [1, 7, 8]. The difference in the light outcoupling efficiency was the reason for the different peak intensities between these two modes. The intermediate peak at wavelength of 535.8 nm, appearing within the stop-band and excited only when the pump light was centred within the mixed-order resonator, was a cavity mode [9], selected by the second-order gratings.

For nanoimprinted polymer lasers, the thresholds previously reported in the literature, to my best knowledge, were above 500 W/cm^2 and in most cases above 1,000 W/cm^2 [10–13]. The sub-hundred W/cm^2 lasing thresholds achieved in this work with both the standard second-order and the mixed-order UV-NIL resonators are therefore the lowest record in the world.

Furthermore, the lasing thresholds of the mixed-order gratings with 4, 8, 15 and 30 second-order periods were found to be very similar. There was no indication of increase in threshold when the number of periods increased from 4 to 30.

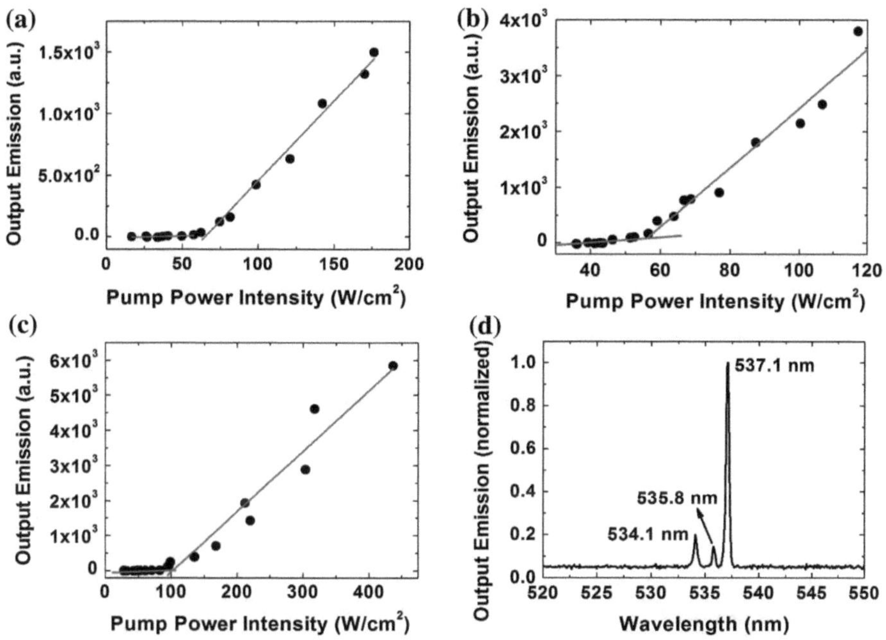

Fig. 6.17 **a** Mixed-order laser pumped with a circular beam; **b** mixed-order laser pumped with a stripe beam; **c** reference laser pumped with a stripe beam; **d** mixed-order laser spectra above threshold

6.8 BBEHP-PPV Laser Pumped by LEDs

To study the prospect for successful micro-LEDs pumping of a polymer laser, the commercial LED was next used as an equivalent pump source in the laser threshold characterization, because it had similar pulse-duration and experimental configuration to the micro-LEDs.

6.8.1 Commercial LEDs Pumping

A 450 nm LUXEON Rebel LED was driven by the laser diode driver PCO-7110-100-7 and produced a maximum optical output power density of at least 750 W/cm^2 with pulse duration of 23 $\pm$ 2 ns from a total light-emitting area of 1.3 by 1.3 mm, see Fig. 6.18a. Figure 6.18b shows the polymer laser with a UV-NIL second-order reference grating pumped by the blue LED at 20 Hz. As the driving current through the LED increased, the emission intensity from the BBEHP-PPV laser showed a linear increase up to an optical excitation intensity of 500 W/cm^2 where a change in the slope of the input–output curve indicating the onset of lasing at 536.8 nm.

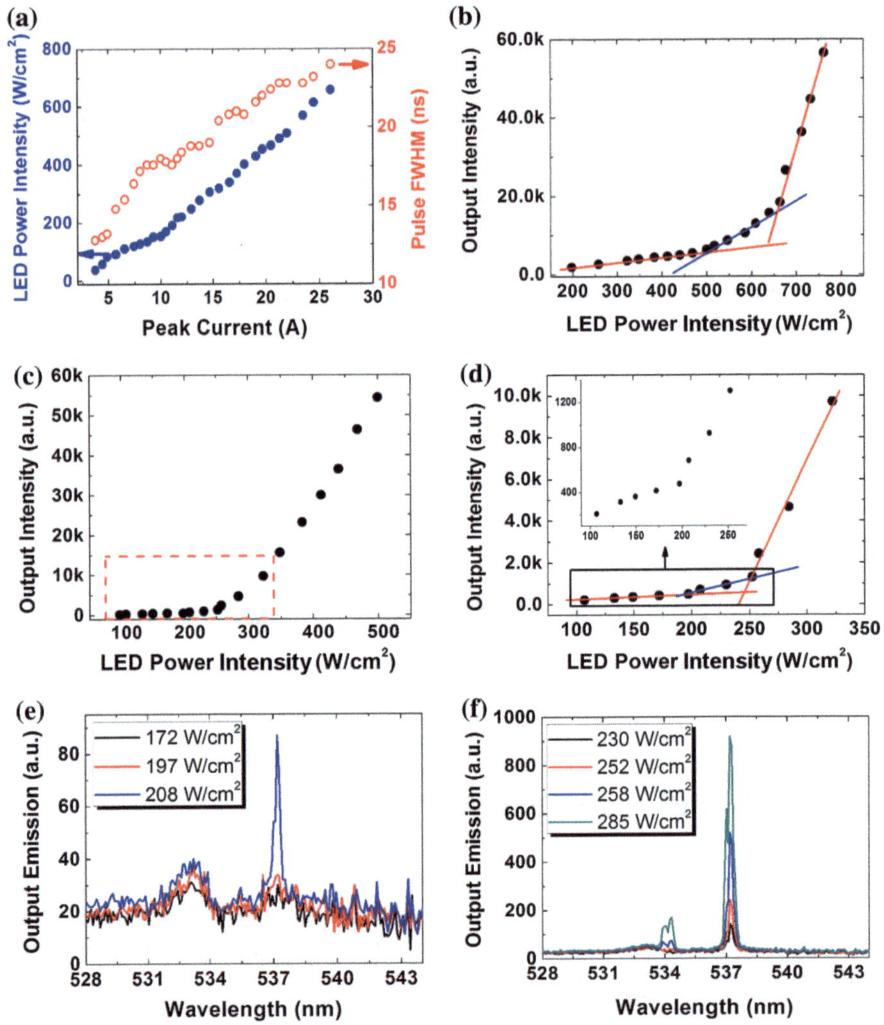

Fig. 6.18 **a** Output intensity (*closed circles*) from a pulsed LUXEON LED as a function of driving current and the corresponding pulse duration (*open circles*); **b** polymer output emission at the wavelength of 536.8 nm with a second-order DFB resonator as a function of input LED intensity; **c** polymer output emission at the wavelength of 537.1 nm with a mixed-order resonator as a function of input LED intensity, and the highlighted region is re-plotted in **d** to show the double thresholds; **e** and **f** are the emission spectrum at different pump intensities showing the onset of two lasing peaks

A higher efficiency of laser emission began when the LED intensity further increased to 650 W/cm^2.

Next a NIL mixed-order DFB laser with 30 periods of second-order gratings for outcoupling was pumped with the same LED. Figure 6.18c and d shows the input–output curve in different LED power ranges with the emission spectra at several

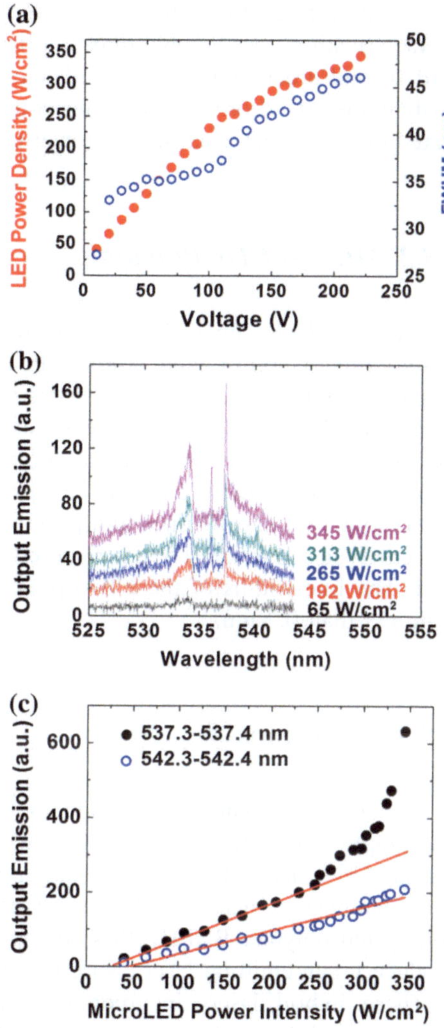

Fig. 6.19 **a** Output power density (*closed circles*) and FWHM of the optical pulses (*open circles*) from the pulsed micro-LED array; **b** Emission spectra from a mixed-order BBEHP-PPV laser at different pump intensities; **c** output intensity as a function of input micro-LED intensity at different wavelength; 537.3 nm is the band-edge lasing wavelength and 542.3 nm is far away from the three potential lasing modes

pump intensities shown in Fig. 6.18e and f. Lasing action was observed at a wavelength of 537.1 nm and at pump power intensity of 200 W/cm^2. When the LED power increased to about 258 W/cm^2, a second band-edge lasing mode at 534.1 nm came above threshold. The reason that the longer-wavelength mode had slightly lower threshold could be that it had less surface output-coupling losses compared to the shorter-wavelength mode, as mentioned in Chap. 2.

This is so far the lowest threshold of an LED pumped organic semiconductor laser. The result was also found to be repeatable with the UV-nanoimprint lithography fabrication. The threshold reduction in this work is very important as it opens up the opportunity for micro-LED pumping of a nanoimprinted laser to make a compact and convenient laser source.

An interesting observation here was that, in the mixed-order laser threshold characterization, the "double-threshold triangle" is smaller than that in the standard second-order grating. This indicates that the gain overcome the loss more easily in the mixed-order cavity. This could be due to a shorter build-up time in the mixed-order resonator caused by lower cavity loss.

6.8.2 Micro-LEDs Pumping

The 1×0.27 mm micro-LED stripe in device LS-450-10 was used to pump mixed-order nanoimprinted lasers with 30 periods of second-order gratings for output coupling. Figure 6.19 shows a micro-LED array pumped BBEHP-PPV laser—the change in emission spectra and output intensity with excitation intensity. The slope in the input–output curve at 537 nm changed when the intensity from the micro-LED array was increased up to 250 W/cm^2, along with an observation of a narrow linewidth peak appearing at the wavelength of 537.3 nm. The cavity lasing mode at 536 nm started lasing at a higher pump intensity level as well. The line-widths of these two peaks at pump intensity of 345 W/cm^2 were both less than 0.2 nm.

6.9 Conclusions

In conclusion organic semiconductor lasers fabricated by UV-nanoimprint lithography have shown extremely low lasing threshold and been successfully pumped by micro-LED arrays. Nanoimprinted distributed feedback resonators, with feature sizes of sub-hundred nanometres, were combined with a light-emitting conjugated polymer BBEHP-PPV, pumped by custom blue micro-LED arrays to produce green-emitting lasers, with threshold of around 250 W/cm^2.

These hybrid lasers incorporate a simple, reproducible and scalable UV-nanoimprint lithography process, compatible with high-performance micro-LEDs. Therefore they present a compact, low-cost alternative laser source to inorganic diode lasers. This work achieved one of the main goals of the HYPIX project, as the first time that a micro-pixellated GaN LED array pumped the nanoimprinted polymer laser above threshold. It is worth noting that the whole laser system (except the LED driver electronics) was made within the HYPIX project. The compatibility of micro-LEDs with CMOS based control [14], promises hybridized organic semiconductor/gallium nitride/CMOS smart pixel arrays for future applications including lighting and sensing.

References

1. Turnbull, G. A., Yang, Y., Shaw, P., Ruseckas, A., Samuel, I. D. W. (2008) Light-emitting diode pumped polymer lasers. *Proceedings of the Society of Photo-Optical Instrumentation Engineers, 7051.*
2. Gong, Z., Jin, S. R., Chen, Y. J., McKendry, J., Massoubre, D., Watson, I. M., et al. (2010). Size-dependent light output, spectral shift, and self-heating of 400 nm InGaN light-emitting diodes. *Journal of Applied Physics, 107*(1), 013103.
3. Herrnsdorf, J. (2012) *Organic lasers and nanostructured organic films for hybrid integration.* Thesis for the degree of Doctor of Philosophy, University of Strathclyde, Glasgow.
4. Karnutsch, C., Gyrtner, C., Haug, V., Lemmer, U., Farrell, T., Nehls, B. S., et al. (2006). Low threshold blue conjugated polymer lasers with first- and second-order distributed feedback. *Applied Physics Letters, 89*(20), 201108.
5. Zankovych, S., Hoffmann, T., Seekamp, J., Bruch, J. U., & Torres, C. M. S. (2001). Nanoimprint lithography: Challenges and prospects. *Nanotechnology, 12*(2), 91–95.
6. Guo, L. J. (2007). Nanoimprint lithography: Methods and material requirements. *Advanced Materials, 19*(4), 495–513.
7. McGehee, M. D., Diaz-Garcia, M. A., Hide, F., Gupta, R., Miller, E. K., Moses, D., et al. (1998). Semiconducting polymer distributed feedback lasers. *Applied Physics Letters, 72*(13), 1536–1538.
8. Barlow, G. F., Shore, A., Turnbull, G. A., & Samuel, I. D. W. (2004). Design and analysis of a low-threshold polymer circular-grating distributed-feedback laser. *J Opt Soc Am B, 21*(12), 2142–2150.
9. Vasdekis, A. E., Tsiminis, G., Ribierre, J. C., O'Faolain, L., Krauss, T. F., Turnbull, G. A., et al. (2006). Diode pumped distributed Bragg reflector lasers based on a dye-to-polymer energy transfer blend. *Optics Express, 14*(20), 9211–9216.
10. Pisignano, D., Persano, L., Mele, E., Visconti, P., Anni, M., Gigli, G., et al. (2005). First-order imprinted organic distributed feedback lasers. *Synthetic Metals, 153*(1–3), 237–240.
11. Mele, E., Camposeo, A., Stabile, R., Del Carro, P., Di Benedetto, F., Persano, L., et al. (2006). Polymeric distributed feedback lasers by room-temperature nanoimprint lithography. *Applied Physics Letters, 89*(13), 131109.
12. Chen, Y., Li, Z., Zhang, Z., Psaltis, D., & Scherer, A. (2007). Nanoimprinted circular grating distributed feedback dye laser. *Applied Physics Letters, 91*(5), 051109.
13. Namdas, E. B., Tong, M., Ledochowitsch, P., Mednick, S. R., Yuen, J. D., Moses, D., et al. (2009). Low thresholds in polymer lasers on conductive substrates by distributed feedback nanoimprinting: Progress toward electrically pumped plastic lasers. *Advanced Materials, 21*(7), 799–802.
14. McKendry, J. J. D., Rae, B. R., Gong, Z., Muir, K. R., Guilhabert, B., Massoubre, D., et al. (2009). Individually addressable AlInGaN micro-LED arrays with CMOS control and subnanosecond output pulses. *IEEE Photonics Technology Letters, 21*(9–12), 811–813.

Chapter 7
Polymer with Intrinsic Microporosity Used as Explosive Vapour Sensors

7.1 Introduction

Over the last two decades, solution-processable organic semiconductors have attracted a great deal of interest. A wide range of optoelectronic devices with this novel class of materials have been investigated, including organic semiconductor lasers presented in the last few chapters. The combination of great photophysical properties, including high gain and broad spectra, with the possibility of large-scale and low-cost fabrication processes, for example using nanoimprint techniques on flexible substrates, has opened up multiple directions for applications of these plastic lasers including amplifiers [1, 2], spectroscopy [3] and chemosensing [4].

One emerging application of fluorescent polymers is the sensing of explosives, in particular nitroaromatic-based explosives [5–7]. In this chapter, organic polymer films and lasers used for trace explosive detection are investigated. In Sect. 7.2 the three main categories of current techniques for explosive sensing are reviewed. The requirements of a sensor are listed and the sensing mechanism for using fluorescent polymers is explained. In Sect. 7.3 a polymer with intrinsic microporosity is introduced as a potential sensing candidate because of its microporous structure. Following the photophysical and ASE studies of this polymer in Sect. 7.4, the prospect of using it in explosive sensors is investigated, with both thin-films in Sect. 7.6 and DFB lasers in Sect. 7.7. The sensitivity of this microporous polymer is studied and the performances are compared in detail with polyfluorene sensors in Sect. 7.7. The results are then summarised in Sect. 7.8.

7.2 Sensing Techniques and Principles

There is currently a critical need for trace explosive detection due to the increased security issues and threats across the world. High-sensitivity detection techniques are needed to avoid missing potentially dangerous objects during routine checks at airports, government buildings and other places at risk of being targeted by terrorists.

Y. Wang, *Low Threshold Organic Semiconductor Lasers*, Springer Theses,
DOI: 10.1007/978-3-319-01267-4_7, © Springer International Publishing Switzerland 2014

These devices could also be employed in humanitarian demining, i.e. the clearing of landmines left after military action. A range of optical spectrometric and chemical analytical methods have been investigated for this application [8–11].

7.2.1 Main Non-Polymer Sensing Techniques

There are currently many different techniques used to detect explosives. Here we briefly review three techniques mostly used in demining actions—electromagnetic imaging, trained animals and spectrometry.

Electromagnetic imaging is a successfully developed traditional method for detecting explosives both in home security and demining. It uses electromagnetic pulses, consisting of for example ground-penetrating radar (GPR), millimetre wave or X-ray, to image the subsurface [12]. The basic principle behind this technique is that different materials or objects absorb, scatter/backscatter or conduct the radiation differently. Take the GPR for example, by detecting the feedback signals from underground with proper calibrations, an image can be used to infer the presence of bulk explosives, even without the release of molecules from its bulk. GPR can be used in a variety of media, including water, rock, road, pavements and soil, however, it suffers from limited performance in high-conductivity or highly scattering environments, and the portability and mobility of devices using this technique are not great.

Nowadays animals, with a great deal of success, are used to detect hidden explosives. The use of dogs for mine detection has expanded dramatically in the last ten years [13, 14]. Their inherent acute sense of smell allows dogs to be one of the most sensitive demining techniques. African pouched rats are also used as indicators of buried explosives in field conditions relying on their smelling capacity to find food. Pigs [15] and honey bees [16, 17] have also joined the mine detecting team in research projects. Trained animals/insects are able to detect mines using the extremely faint odour of explosives and other chemicals leaking from mines. However, there are problems with the use of animals/insects as explosive detectors. They require time-consuming and expensive training, they tire, their capabilities decline with time, and they can be easily distracted in complicated field environments.

Spectroscopic techniques are concerned with the absorption, emission or scattering of electromagnetic radiation by atoms or molecules [12]. The technique that has been applied most ubiquitously for high-speed detection of trace explosive vapours and gases is ion mobility spectrometry (IMS), mostly because of the availability of portable instruments [10]. In IMS, sample vapours are converted to ions at atmospheric pressure and those ions are characterized by their gas phase mobility spectrum, i.e. a graph of ion current versus time of ion drift, more accurately in weak electric fields [9]. This method is used in detection of explosives with relatively high selectivity and sensitivity. Other various vibrational spectroscopic techniques, including infrared absorption and Raman scattering spectroscopy, are widely used across the world for explosive detection as well [11, 18, 19].

7.2.2 Criteria for an Explosive Sensor

Trinitrotoluene (TNT) is the major explosive component of most landmines. Military-grade TNT includes quantities of several nitroaromatic compounds including 2,4-dinitrotoluene (DNT), 2,6-DNT, 1,3-dinitrobenzene (also known as *m*DNB), and 1,4-DNB (also known as *p*DNB), and these molecules generate higher vapour concentrations than 2,4,6-TNT itself [20], as shown in Fig. 7.1. TNT has a saturated vapour pressure of $\sim$10 part-per-billion (ppb) at room temperature [21].

For sensing explosive vapours released from buried suspicious objects (taking TNT as an example), the sensors must fulfil a number of criteria [22]. The first one is sensitivity, that is, a distinguishing signal resulting from the presence of TNT vapour with a concentration of a few parts per billion (ppb) and below; second is stability, i.e. the sensor must work reliably in different and variable environments; third is selectivity, discrimination of those low concentrations of explosives from much higher concentrations of other chemical species, and thus ensuring low false alarm rate; and fourth is a capability for real-time signal processing. The final one, not essential but desirable, is reversibility, by removing the vapour from the sensor, the signal should return to its initial state within a short period of time. For in-field applications, a portable system would also be advantageous. Moreover, the capability to detect a large range of explosive molecules and at the same time with good selectivity would be ideal. Research in this field continues to improve detection limits, speed of response, selectivity and portability [7, 23].

7.2.3 Sensing Mechanism in Luminescent Polymers

Most explosives consist of organic compounds containing nitro or nitroaromatic groups, as shown in Fig. 7.1. The highly nitrated molecules in these explosives make them electron-deficient. When these electron-acceptors are in contact with electron-rich semiconducting polymers, photo-excited electrons in the polymer can transfer from the lowest unoccupied molecular orbital (LUMO) of the polymer to the LUMO of a nearby analyte molecule with a lower energy level. Normally

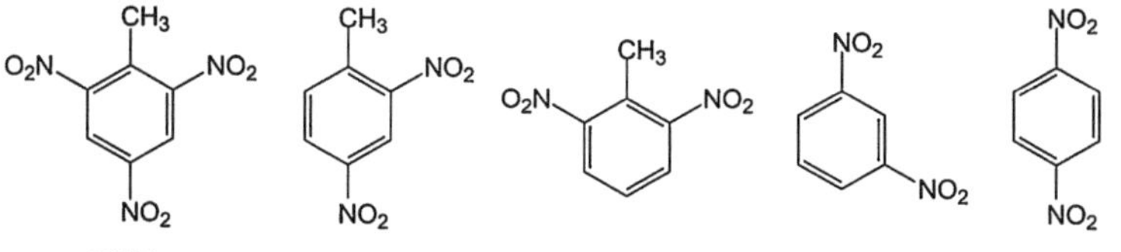

	TNT	2,4-DNT	2,6-DNT	1,3-DNB	1,4-DNB
VP	1	18	70	110	3

Fig. 7.1 The molecular structures of the commonly used explosive compounds and their relative vapour pressures (VP) against 2,4,6-TNT which is 10 ppb at 25 °C

the photo-excited electron and hole in the polymer's singlet state would recombine to emit a photon, as described in Chap. 2. However, the charge transfer between the polymer and explosives can compete with a non-radiative recombination and cause efficient quenching of the fluorescence, as shown schematically in Fig. 7.2.

The world's first fluorescent polymer films used as TNT sensors, were demonstrated in 1998 by Yang and Swager [21]. A group of fluorescent pentiptycene-derived polymers with rigid three-dimensional backbones were designed to be highly sensitive to TNT vapour, as the backbones prevented self-quenching of the fluorescence in the polymer film and at the same time provided porosity in the thin-film for efficient analyte binding.

In 2005, Rose and co-workers reported an organic microring laser providing high sensitivities to nitroaromatic analyte [4]. By using a polymer laser, 5 ppb TNT was detected by observation of attenuation in the lasing emission before any change of the fluorescence could be measured. However, a microring is an inconvenient laser geometry, as it emits light in all directions rather than as a collimated beam. Subsequent work has explored other organic semiconductor laser sensors with conventional DFB resonator configurations [24, 25]. Sensing enhancements with stimulated emission have been observed, in comparison with the photoluminescence sensing scheme using thin films, which provides an alternative way for trace explosive detection [26].

In light-emitting polymers, the photoluminescence can be dramatically influenced by small numbers of binding events, i.e. a single explosive molecule can quench many repeat units of a polymer chain. In addition, neighbouring polymer chains can also be quenched as in the film, interchain exciton transfer enables excitons formed on one chain to move to quenching sites on other chains to some extent [5]. A disadvantage of a solid film is that diffusion of the analyte into the film can be very slow resulting in the speed of response being limited and the sensor becoming easy to saturate.

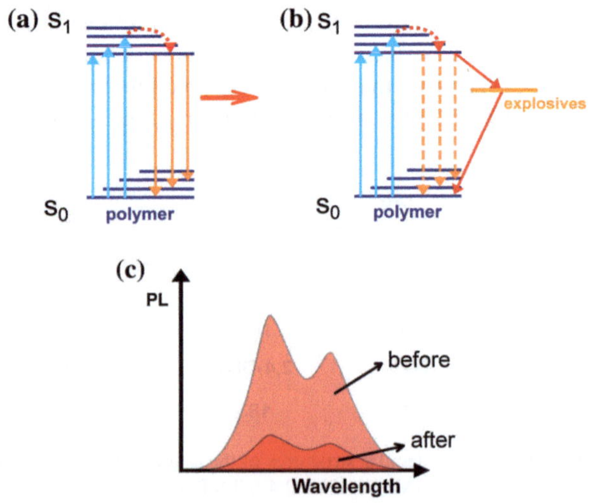

Fig. 7.2 Schematic drawing of energy levels and electron transfer for polymer **(a)** without and **(b)** with the presence of explosive molecules; **c** schematic of quenching of the fluorescence in the polymer after exposure to explosive molecules

7.3 Polymer with Intrinsic Microporosity

To address the issue of analyte diffusion, I explored the effect of using a microporous polymer—a solution-processable polymer with an intrinsic micropore structure, polybenzodioxane PIM-1. The polymer used in this work was synthesised by Prof. Neil McKeown and Dr Kadhum Msayib at Cardiff University [27, 28].

The intrinsic microporosity of the polymer forms as a consequence of the rigidity of the macromolecular chain and its contorted structure. The polymer molecule consists of an alternation of a rigid fused ring structure and twisted spiro linkages. The rigid and twisted backbone bonding with little rotational freedom is essential to create well-defined cavities and to prevent collapse of the micropores [29], see Fig. 7.3. This polymer with intrinsic microporosity (PIM-1) and related PIMs have been used in the past as robust gas separation membranes [30–32], trace organic volatiles indicators[33] and pre-concentrator traps for explosive vapours [34].

A convenient way to use organic polymers to detect explosive vapour is to measure the change in fluorescence intensity of an optically excited thin film. Amplified spontaneous emission (ASE) was later found to show a much greater change than PL [4]. To investigate the potential of PIM-1 as a light-emitting material for sensing trace vapours of nitro-aromatic molecules, the photophysical properties and ASE were studied first.

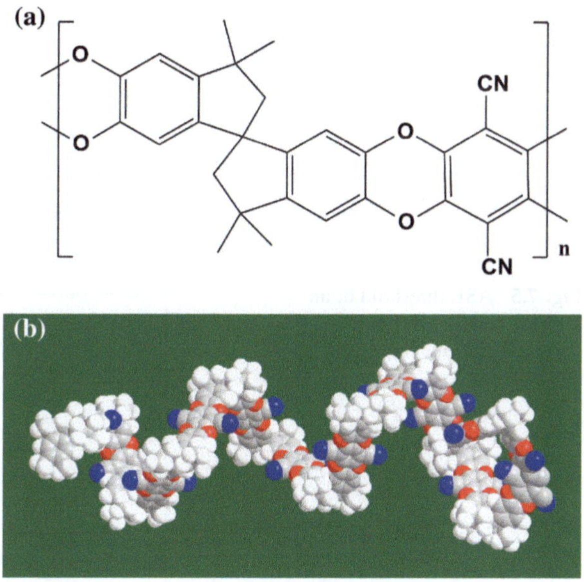

Fig. 7.3 **a** Molecular structure of PIM-1; **b** 3D model of part of a rigid and contorted polymer chain of PIM-1, created with ChemDraw 3D

7.4 Photophysical and ASE Studies of PIM-1

A 340 nm thick PIM-1 film was spin-coated from a 30 mg/ml chloroform solution onto a silica substrate. Absorption and PL spectra were recorded with spectrophotometers, and the PLQY of the thin films were measured using the Hamamatsu measurement system. Fig. 7.4 shows the absorption and PL (excited at 400 nm) spectra from the 340 nm thick PIM-1 film. The absorption and emission spectra have peaks at 416 nm and 501 nm, respectively. The solid-state PLQY was measured to be 21 ± 2 %.

In order to assess the optical gain properties of PIM-1, ASE detected from the edge of the films was studied. 4 ns pulses from the OPO at a wavelength of 425 nm were used as the excitation source, with a repetition rate of 20 Hz. A cylindrical lens was used to shape the pump beam into a 3.2 × 0.17 mm stripe on the surface of the polymer films. The characteristic spectral narrowing of the edge-emission light was observed with an emission peak wavelength at 528 nm. Fig. 7.5 shows the linewidth of the edge-emission spectrum and relative output intensity as a

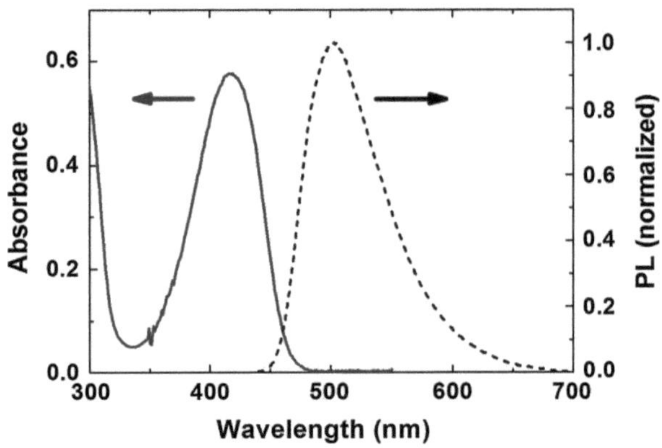

Fig. 7.4 Absorption (*solid line*) and PL spectra (*dashed line*, with excitation wavelength of 400 nm) of a 340 nm thick PIM-1 film

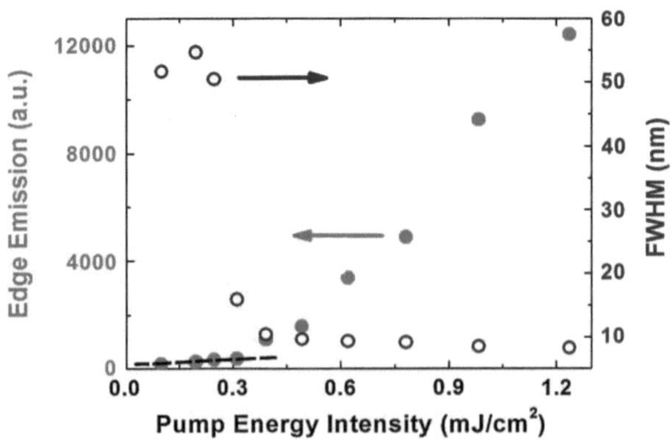

Fig. 7.5 ASE threshold of an air-PIM-silica waveguide; polymer film thickness is 340 nm

function of the pump energy intensity. The linewidth dropped dramatically from 52 to 8 nm when pumping above a threshold intensity of 310 μJ/cm^2 [35]. This relatively high ASE threshold was expected with its low solid-state PLQY value.

7.5 Set-Up of Sensing Experiment

The nitroaromatic analyte used in this investigation was 1,4-DNB. In the PL sensing experiments, the PIM-1 thin film was placed in an air-tight optical chamber and excited by a pulsed frequency tripled Nd:YVO_4 microchip laser at 355 nm at 1 kHz repetition rate. The PL emission spectrum from the film was monitored with the fibre-coupled CCD detector. The optical chamber was connected to a valve-controlled gas port with two gas supplies allowing either DNB vapour or clean nitrogen flow through the chamber, as shown in Fig. 7.6. An exhaust port from the chamber was connected to a solvent bubbler then to the exhaust pipe. Saturated vapour pressure of 1,4-DNB at room temperature is about 30 ppb, as shown in Fig. 7.1. By controlling the nitrogen pressure at 2 bar, this configuration provides a flow with $\sim$10 ppb 1,4-DNB by volume [25].

This set-up was designed to allow measurement of the dynamic response of the PL sensing process. The chamber was first purged with clean nitrogen then the gas port was switched to the flow with a contamination of 10 ppb DNB. For continuous exposure experiments the dynamics of the PL emission were monitored for a time period of a few minutes.

Additionally with this set-up, a cumulative PL sensing scheme can be achieved by monitoring the PL emission while switching the input-gas flow between clean nitrogen and DNB contaminated nitrogen—alternating the chamber atmosphere between low ($\sim$0 ppb) and "high" ($\sim$10 ppb) DNB vapour.

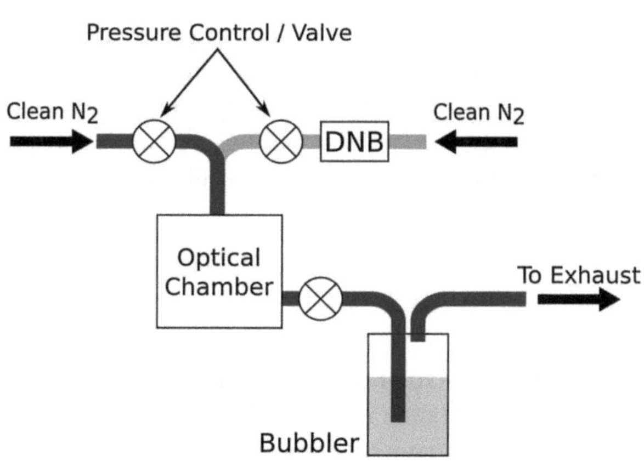

Fig. 7.6 Schematic of the sensing set-up

7.6 PL Sensing Using PIM-1 Films

7.6.1 Continuous Sensing

When a 62 ± 10 nm thick PIM-1 film, excited at 355 nm, was exposed to 10 ppb DNB vapour, the PL was quenched rapidly. The dynamics of the PL quenching are shown in Fig. 7.7. Prior to DNB exposure, during a 3-min nitrogen purge, the emission from the film was fairly stable suggesting that there was very little photo-induced degradation caused by UV excitation. Once the film was exposed to the DNB vapour the emission reduced instantly.

After a 30-sec exposure, the emission intensity of the 134 nm thick film was reduced by 10 ± 2 %, see Fig. 7.7c, and after 3-min, the emission intensity was half the initial value. After a 30-min continuous exposure to DNB vapour, the film emission intensity was reduced by over 90 %. A similar change in the PLQY of the PIM-1 film was also observed, decreasing from 21 % (pre-exposure) to 3 % (post-exposure).

The sensing efficiency, defined as the ratio of the change of the emission intensity during the exposure period to the initial PL intensity,

$$Efficiency(t) = \frac{I(t = 0) - I(t)}{I(t = 0)} \times 100\,\% = (1 - \frac{I(t)}{I(t = 0)}) \times 100\,\% \qquad (7.1)$$

was increasing as the exposure time increased and stabilized after approximately 8 min for the 62 nm thick film, indicating the saturation of the sensor. As seen in Fig. 7.7, the sensing efficiency of the thicker film (134 nm) saturated after a longer period of time, about 25 min, of continuous exposure to 10 ppb DNB vapour. This suggests that the explosive molecules required a small amount of time to penetrate through the polymer film to quench more fluorophores. In Fig. 7.7c, we can see that the thinner film has a slightly faster initial response than the thicker film, which is also due to the difference in film thicknesses.

With a 340 ± 20 nm thick PIM-1 film, the PL emission intensity was reduced by 31 % after a 3-min exposure to 10 ppb DNB vapour. Subsequent to a 30-min continuous exposure, the sensing efficiency was stabilized at 85 %. The sensing efficiency of the thick film was lower than for the thinner films above, but it required a longer time to reach saturation. Therefore film thickness is a very important parameter in PL intensity sensing. Too thin or too thick may affect both the sensitivity, speed of response and saturation limit of the sensor. More film-thickness dependent PL sensing results will be discussed in Chap. 8.

7.6.2 Cumulative Sensing

I therefore next explored whether this strong binding could be used to make a sensor with an accumulating response that may be useful for applications in shipping and storing of sensitive items or in the measurement of total exposure time.

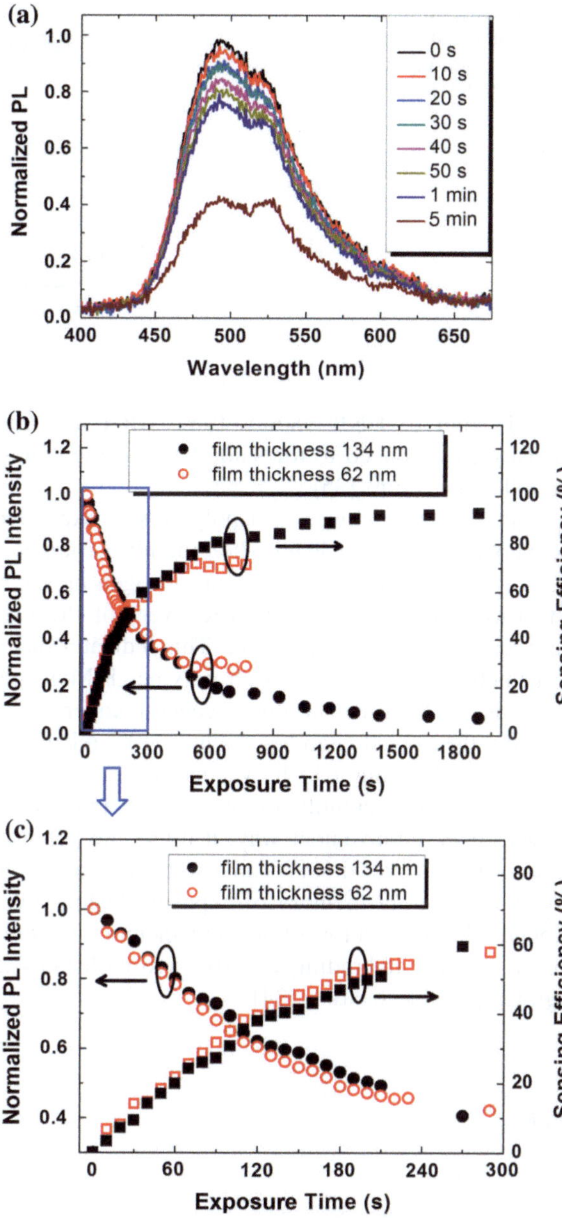

Fig. 7.7 **a** PL spectra of a 62 nm thick PIM-1 film during 5 min exposure to DNB; **b** dynamics of PL sensing with a 62 nm (*red*) and 134 nm (*black*) thick PIM-1 films; **c** first 300 s of the quenching dynamics, highlighted in **b**

In the cumulative PL sensing, a 340 nm thick PIM-1 film was first exposed to clean nitrogen for 60 s, and then exposed to DNB vapour for 60 s resulting in an approximately 10 % drop in the film PL intensity, as shown in Fig. 7.8. In the following 1-min clean nitrogen purge, the emission intensity remained unchanged. Further quenching of the PL emission resumed when the film was exposed to a

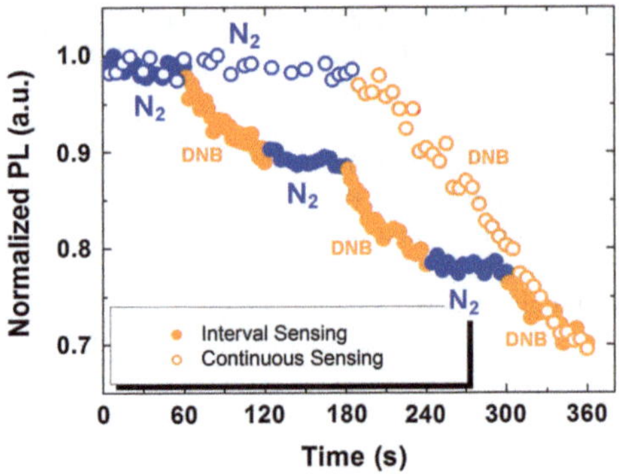

Fig. 7.8 Interval (*closed circles*) and continuous (*open circles*) PL sensing with a 340 nm PIM-1 film

second-round of DNB vapour flow. After a 3-min exposure in total, the PL intensity was reduced to the same level as the continuous 3-min exposure with a same-thickness film [35].

Normally using fluorescent polymers in explosive sensors, the light emission recovers to its original level when the nitroaromatic molecules are removed from the polymer film by flowing clean nitrogen or vacuum-pumping the system [5]. When using PIM-1 however no recovery of PL was observed when the vapours were removed from the chamber. This indicates that extremely strong interactions occur between the electron-rich polymer PIM-1 and the electron-deficient DNB molecules, preventing the fluorescence recovery. This was confirmed by vacuum pumping the optical chamber down to 10^{-4} mbar. No increase in the PL intensity was observed during a 15-h evacuation.

Although the accumulation of the DNB in the PIM-1 film prevents the reuse of the sensor in a convenient way, it offers the potential to sense very low concentration vapours over time; it can also be used as an efficient explosives absorber. It allows an inherent memory of exposure history for the polymer PIM-1. Fig. 7.9 explains the application of sensing memory vividly. Moreover, it has been discussed that the unbinding of explosive molecules to PIM–1 can be realised by high-temperature heating [34].

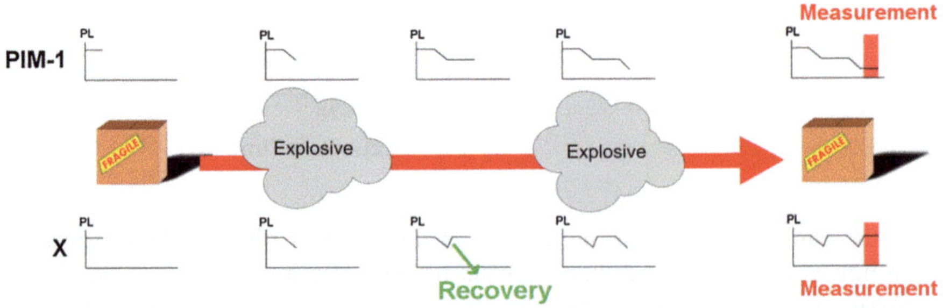

Fig. 7.9 Accumulation of the explosives in PIM-1 records a sensing history; X is a recoverable polymer, for example PFO, which in this case cannot indicate the presence of explosives

7.7 Highly Sensitive Laser Sensor

Stimulated emission from the polymer has been shown to allow enhancement of the detected signals, which is critical for sensing trace amounts of explosives within extremely short times [4, 24, 25], as mentioned in Sect. 7.2.3. However, data on response time is so far extremely limited. The feasibility of making a laser from PIM-1 and using it for explosive detection is further explored in this section.

7.7.1 PIM-1 Laser Sensor

The refractive index of PIM-1 films was measured using the variable angle spectroscopic ellipsometer. The effective refractive index of an air-polymer-SiO_2 waveguide was calculated to determine the optimal distributed feedback (DFB) laser configuration.

The polymer films were found to be isotropic when spin-coated. The refractive index of PIM-1 at 528 nm (the peak of ASE) is 1.59. This value is rather low compared to conventional conjugated polymers, which reflects the low density of the PIM-1 films due to its microporous structure. The effective refractive index of the guided TE_0 mode was calculated to be 1.49 for a 340 nm thick waveguide.

A surface-emitting PIM-1 DFB laser was fabricated with a 1D silica grating with a period of 350 nm, and a gain medium thickness of 340 $\pm$ 20 nm. The PIM-1 laser was optically pumped with 425 nm light from the OPO. Above a pump threshold of 174 $\mu J/cm^2$, the laser emitted a narrow wavelength peak at 529 nm [35].

The PIM-1 laser was then immersed in a nitrogen flow with ~10 ppb DNB. After a 5-min exposure the laser threshold increased to 436 $\mu J/cm^2$, see Fig. 7.10. The post-exposure laser threshold was 2.5 times higher than the pre-exposure threshold. A reduction of 4.4 times in the slope efficiency was observed at the same time.

Fig. 7.10 PIM-1 Laser threshold measured before and after 5-min exposure to 10 ppb DNB

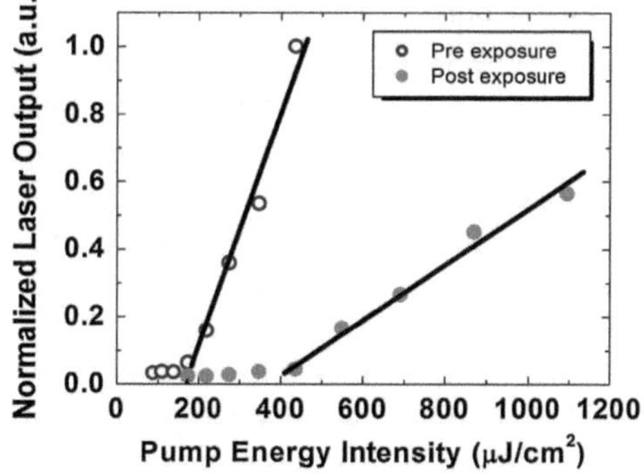

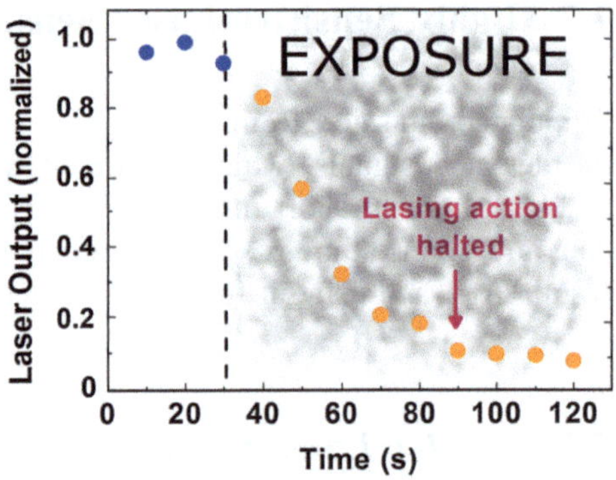

Fig. 7.11 PIM-1 Laser emission dynamics during a 90 s exposure when the laser is continuously excited at an energy intensity of 346 μJ/cm^2

Fig. 7.11 shows the dynamics of the laser emission during a 90 s exposure to the DNB vapour with the excitation energy set to twice the pre-exposed laser threshold, 346 μJ/cm^2. The PIM-1 laser output intensity decreased rapidly by 60 % within the first 30 s. Laser action switched off after 60 s exposure, due to the increase of the threshold, and the decrease of emission intensity slowing down when only spontaneous emission was observed from the polymer film. No indication of recovery of laser emission was observed with these laser sensors when the explosive vapour was removed.

As mentioned in Sect. 7.6, the PL sensing efficiency of a 340 nm thick film in a 60 s exposure is approximately 10 %. In the laser sensor, the sensing efficiency is 63 ± 5 % within 30 s and almost 90 % within 60 s. The very convenient DFB laser geometry has a significantly enhanced sensing efficiency and much faster responsivity in the low vapour pressure explosives detection.

7.7.2 Comparison with PFO Sensors

Compared to a linear conjugated polymer, polyfluorene (PFO), the polymer of intrinsic microporosity (PIM-1) with similar film thickness has shown further advantages. Figure 7.12 presents the PL sensing dynamics of PIM-1 and PFO films within a 900s (15 min) exposure period [26]. It is obvious that the PL from a porous film is quenched faster than from a dense non-porous film, and the sensing efficiency is also much lower in the latter film.

The PIM-1 laser sensor also has a higher sensing efficiency and a much faster responsivity compared to the PFO laser sensor. After 5-min exposure to 10 ppb DNB vapour, the pump threshold of the PFO laser increased by 1.8 times, while the PIM-1 laser threshold increased by 2.5 times with a larger change in the slope efficiency [26, 35].

The rapid responsivity and high sensitivity of the PIM-1, summarised in Table 7.1, is due to its microporous structure and increased surface area, which can increase the penetration rate of the analyte molecules into the film (see

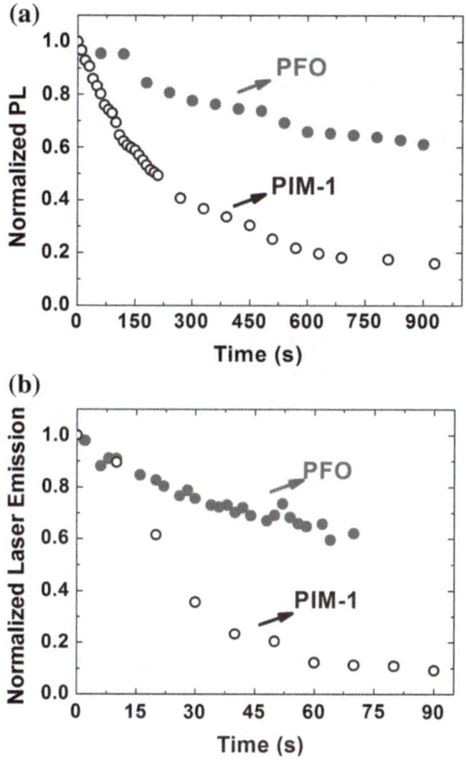

Fig. 7.12 Comparison between PFO and PIM-1 sensing dynamics in **a** PL and **b** DFB laser

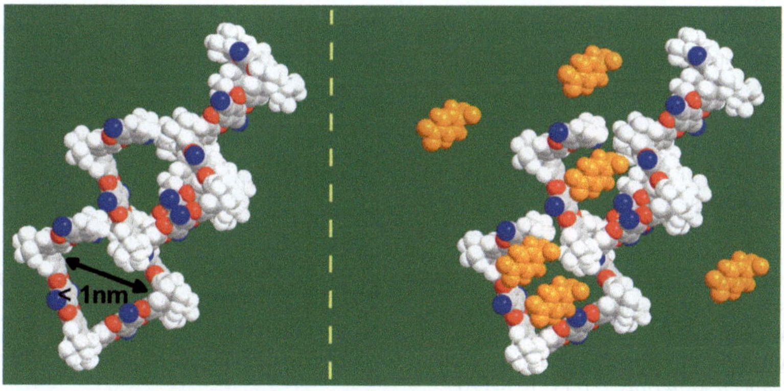

Fig. 7.13 3D models of PIM-1 chain (*left*) without and (*right*) with 1,4-DNB molecules filling the sub-nanometre sized pores

Table 7.1 Comparison of sensing parameters between PIM–1 and PFO both exposed to ~10 ppb DNB vapour

Sensing characteristics	PIM-1	PFO
PL sensing efficiency (15 min)	82 %	40 %
Increase in laser threshold (5 min)	2.5 ×	1.8 ×
Decrease in slope efficiency (5 min)	4.4 ×	3 ×
Maximum sensing efficiency (5 min)	95 %	82 %
Relative response speed (decrease in laser output intensity)	60 % in 30 s	40 % in 46 s

Fig. 7.13). The pore size distribution of PIM-1 is mostly sub-nanometre with a maximum of 0.65 nm, determined using the Horvath-Kawazoe analysis of low pressure nitrogen adsorption at 77 K [30–32]. At the same time, the strong binding between the DNB molecules and PIM-1 are expected to assist the sensing to some extent.

7.8 Conclusions

In this chapter, the polymer PIM-1 has been found to be a good indicator for low vapour pressure (~10 ppb) explosives. The presence of the analyte, 1,4-DNB, can be indicated by monitoring the photoluminescence from thin-films. The response accumulates suggesting a very strong interaction between DNB and the microporous polymer, and leading to an inherent memory ability, which has the potential for applications requiring an integrated measurement of vapour concentration.

The PIM-1 laser has shown a significantly enhanced sensing efficiency and a much faster responsivity in low vapour pressure explosives detection, and potentially can be used as an ultra-sensitive sensor. As the DFB lasers can be readily fabricated by nanoimprint, and as they can also be pumped by LEDs, there is the prospect of low-cost, portable explosive sensors based on light-emitting polymers.

References

1. Lawrence, J. R., Turnbull, G. A., & Samuel, I. D. W. (2002). Broadband optical amplifier based on a conjugated polymer. *Applied Physics Letters, 80*(17), 3036–3038.
2. Amarasinghe, D., Ruseckas, A., Vasdekis, A. E., Turnbull, G. A., & Samuel, I. D. W. (2009). High-gain broadband solid-state optical amplifier using a semiconducting copolymer. *Advanced Materials, 21*(1), 107–110.
3. Woggon, T., Klinkhammer, S., & Lemmer, U. (2010). Compact spectroscopy system based on tunable organic semiconductor lasers. *Applied Physics B, 99*(1–2), 47–51.
4. Rose, A., Zhu, Z. G., Madigan, C. F., Swager, T. M., & Bulovic, V. (2005). Sensitivity gains in chemosensing by lasing action in organic polymers. *Nature, 434*(7035), 876–879.

5. Thomas, S. W., Joly, G. D., & Swager, T. M. (2007). Chemical sensors based on amplifying fluorescent conjugated polymers. *Chemical Reviews, 107*(4), 1339–1386.
6. Germain, M. E., & Knapp, M. J. (2009). Optical explosives detection: From color changes to fluorescence turn-on. *Chemical Society Reviews, 38*(9), 2543–2555.
7. Salinas, Y., Martinez-Manez, R., Marcos, M. D., Sancenon, F., Costero, A. M., Parra, M., et al. (2012). Optical chemosensors and reagents to detect explosives. *Chemical Society Reviews, 41*(3), 1261–1296.
8. Kercel, S. W., Burlage, R. S., Patek, D. R., Smith, C. M., Hibbs, A. D., & Rayner, T. J. (1997). Novel methods for detecting buried explosive devices. *P Soc Photo-Opt Ins, 3079*, 467–477.
9. Ewing, R. G., Atkinson, D. A., Eiceman, G. A., & Ewing, G. J. (2001). A critical review of ion mobility spectrometry for the detection of explosives and explosive related compounds. *Talanta, 54*(3), 515–529.
10. Moore, D. S. (2004). Instrumentation for trace detection of high explosives. *Review of Scientific Instruments, 75*(8), 2499–2512.
11. Steinfeld, J. I., & Wormhoudt, J. (1998). Explosives detection: a challenge for physical chemistry. *Annual Review of Physical Chemistry, 49*, 203–232.
12. WOODFIN, R. L.,(2007). *Trace chemical sensing of explosives*. wiley-Interscience: .
13. Schoon, A., & Berntsen, T. G. (2011). Evaluating the effect of early neurological stimulation on the development and training of mine detection dogs. *Journal of Veterinary Behaviour, 6*(2), 150–157.
14. McLean, I. G. (2003). *Mine Detection Dogs: Training, Operations and Odour Detection*. Geneva: Geneva International Centre for Humanitarian Demining (GICHD).
15. Townsend, J. (2003). Pigs, a demining tool of the future? *J. Journal of Mine Action, 7*(3), 43.
16. Shaw, J. A., Seldomridge, N. L., Dunkle, D. L., Nugent, P. W., Spangler, L. H., Bromenshenk, J. J., et al. (2005). Polarization lidar measurements of honey bees in flight for locating land mines. *Optics Express, 13*(15), 5853–5863.
17. Bromenshenk, J. J.; Henderson, C. B.; Seccomb, R. A.; Rice, S. D.; Etter, R. T.; Bender, S. F. A.; Rodacy, P. J.; Shaw, J. A.; Seldomridge, N. L.; Spangler, L. H.; Wilson, J. J. (2003). Can honey bees assist in area reduction and landmine detection? *Journal of Mine Action, 7* (3).
18. Todd, M. W., Provencal, R. A., Owano, T. G., Paldus, B. A., Kachanov, A., Vodopyanov, K. L., et al. (2002). Application of mid-infrared cavity-ringdown ectroscopy to trace explosives vapor detection using a broadly tunable (6–8 mu m) optical parametric oscillator. *Applied Physics B, 75*(2–3), 367–376.
19. Sylvia, J. M., Janni, J. A., Klein, J. D., & Spencer, K. M. (2000). Surface-enhanced Raman detection of 1,4-dinitrotoluene impurity vapor as a marker to locate landmines. *Analytical Chemistry, 72*(23), 5834–5840.
20. Turk, A. S., & Hocaoglu, K. A. (1977). *Vertiy*. A. A.: Waley & Sons.
21. Yang, J. S., & Swager, T. M. (1998). Fluorescent porous polymer films as TNT chemosensors: electronic and structural effects. *Journal of the American Chemical Society, 120*(46), 11864–11873.
22. Czarnik, A. W. (1998). A sense for landmines. *Nature, 394*(6692), 417–418.
23. Caygill, J. S., Davis, F., & Higson, S. P. J. (2012). Current trends in explosive detection techniques. *Talanta, 88*, 14–29.
24. Richardson, S., Barcena, H. S., Turnbull, G. A., Burn, P. L., & Samuel, I. D. W. (2009). Chemosensing of 1, 4-dinitrobenzene using bisfluorene dendrimer distributed feedback lasers. *Applied Physics Letters, 95*(6), 063305.
25. Yang, Y., Turnbull, G. A., & Samuel, I. D. W. (2010). Sensitive explosive vapor detection with polyfluorene lasers. *Advanced Functional Materials, 20*(13), 2093–2097.
26. Wang, Y., Yang, Y., Turnbull, G. A., & Samuel, I. D. W. (2012). Explosive sensing using polymer lasers. *Molecular Crystals and Liquid Crystals, 554*, 103–110.

27. Budd, P. M., Elabas, E. S., Ghanem, B. S., Makhseed, S., McKeown, N. B., Msayib, K. J., et al. (2004). Solution-processed, organophilic membrane derived from a polymer of intrinsic microporosity. *Advanced Materials, 16*(5), 456–458.
28. Budd, P. M., Ghanem, B. S., Makhseed, S., McKeown, N. B., Msayib, K. J., & Tattershall, C. E. (2004). Polymers of intrinsic microporosity (PIMs): robust, solution-processable, organic nanoporous materials. *Chemical Communications, 2*, 230–231.
29. McKeown, N. B., & Budd, P. M. (2010). Exploitation of intrinsic microporosity in polymer-based materials. *Macromolecules, 43*(12), 5163–5176.
30. Budd, P. M., Msayib, K. J., Tattershall, C. E., Ghanem, B. S., Reynolds, K. J., McKeown, N. B., et al. (2005). Gas separation membranes from polymers of intrinsic microporosity. *Journal of Membrane Science, 251*(1–2), 263–269.
31. Ghanem, B. S., McKeown, N. B., Budd, P. M., Selbie, J. D., & Fritsch, D. (2008). High-performance membranes from polyimides with intrinsic microporosity. *Advanced Materials, 20*(14), 2766.
32. Budd, P. M., McKeown, N. B., Ghanem, B. S., Msayib, K. J., Fritsch, D., Starannikova, L., et al. (2008). Gas permeation parameters and other physicochemical properties of a polymer of intrinsic microporosity: polybenzodioxane PIM-1. *Journal of Membrane Science, 325*(2), 851–860.
33. Rakow, N. A., Wendland, M. S., Trend, J. E., Poirier, R. J., Paolucci, D. M., Maki, S. P., et al. (2010). Visual Indicator for trace organic volatiles. *Langmuir, 26*(6), 3767–3770.
34. Combes, D. J., Cox, T. I., Sage, I. C.(2010). Preconcentrator device incorporating a polymer of intrinsic microporosity. U.S. Patent 2010/0144049.
35. Wang, Y., McKeown, N. B., Msayib, K. J., Turnbull, G. A., & Samuel, I. D. W. (2011). Laser chemosensor with rapid responsivity and inherent memory based on a polymer of intrinsic microporosity. *Sensors-Basel, 11*(3), 2478–2487.

Chapter 8
Towards Ultra-Portable Hybrid Organic/ Inorganic Explosives Sensing Devices

8.1 Introduction

As discussed in Chap. 7, the main criteria for a practical explosive sensor include sensitivity, selectivity, portability, reversibility and fast signal processing. To the best of my knowledge, the only product that uses fluorescent organic polymers as their sensing platform commercially available at market is a series of portable trace explosives detectors, called "Fido" produced by FLIR, which are able to detect common military-grade compounds, homemade explosives and improvised explosive devices [1]. Figure 8.1 shows the general configuration of the key sensing component in most Fido products [1, 2]. The configuration illustrated is built around a glass capillary whose inner surface is coated with a specially designed amplifying fluorescent polymer material. The device is configured in such a way that the system pump pulls air from the environment through the capillary which is gently heated to improve system response. The principle of sensing in Fido is simply based on the quenching of PL intensity from the polymer.

To explore the prospect of portable and low-cost polymer based chemo-sensors used to detect trace nitroaromatic-based explosives, two organic/inorganic hybrid sensing devices have been investigated in this chapter, including a highly compact CMOS/GaN/polymer sensor using PL lifetime as a sensing mechanism in Sect. 8.2 and an integrated LED pumped polymer laser sensor in Sect. 8.3.

8.2 Explosive Sensor Based on a Hybrid Lifetime Micro-System

8.2.1 Sensing Based on Fluorescence Lifetime

Most fluorescence-based sensors using organic semiconductors measure a reduction in the output intensity of the light emission, as described in Chap. 7. An alternative approach is to measure a change in the PL decay lifetime. The quenching of the

Y. Wang, *Low Threshold Organic Semiconductor Lasers*, Springer Theses,
DOI: 10.1007/978-3-319-01267-4_8,

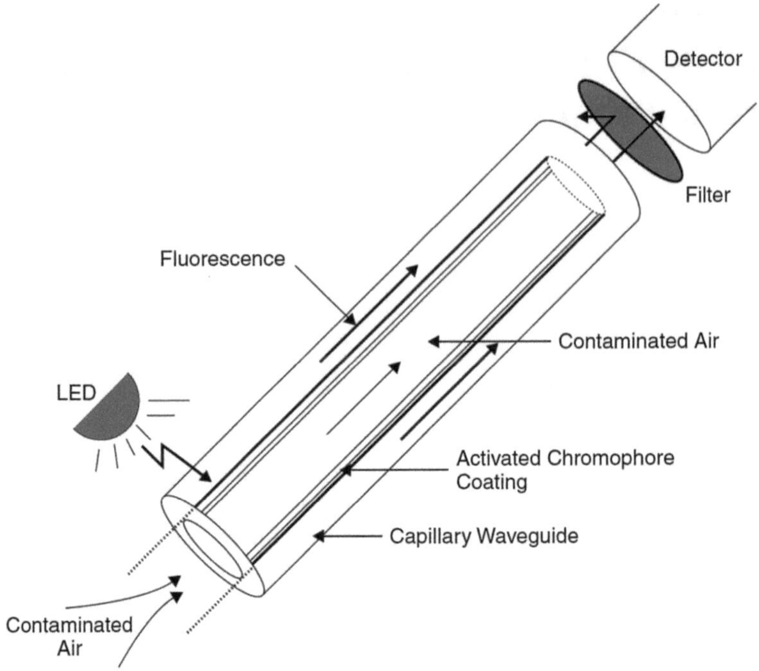

Fig. 8.1 Illustration of the main part of Fido explosives detector; adapted from Ref. [1]

fluorescence increases the non-radiative relaxation of the optically excited polymer, i.e. the polymer's PL lifetime as well as the luminescence intensity decreases as a result of quenching, see Fig. 8.2. Therefore, measuring these decay dynamics provide another way to indicate the presence of quenchers, in this scenario the explosive molecules. Sensors based on measuring the polymer's excited-state lifetime are more readily quantifiable compared to just monitoring the reduction in PL intensity, because the PL decay lifetime is less affected by internal and external environmental factors, including the number of excited states in the polymer, excitation intensity, collection efficiency and stray light.

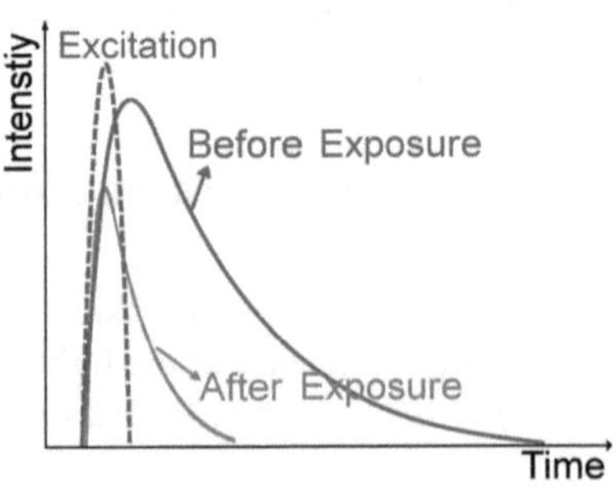

Fig. 8.2 Illustration of explosive sensing based on PL lifetime

Time-resolved fluorescence lifetime measurements have been recently reported for detecting nitro-aromatic molecules dissolved in solution using conjugated dendrimers and polymers [3, 4]. As explained in Chap. 7, using thin films enables a quenching enhancement, as a single analyte molecule is able to quench many chromophores not only on the same but also on the neighbouring polymer chains. In this chapter, detection of our model explosive, 1,4-DNB, at a low vapour pressure, using thin films of a fluorescent conjugated copolymer combined with different time-resolved fluorescence lifetime systems are investigated [5]. The fluorescent conjugated copolymer used here is a proprietary green-light-emitting fluorene based copolymer “CDT Green (CDTG)” supplied by Cambridge Display Technology Limited (CDT Ltd).

8.2.2 Copolymer “CDT Green”

The green-emitting copolymer was expected to have high-yield photoluminescence efficiency. The absorption and PL spectra of a 250 nm thick CDTG polymer film, spin-coated onto silica substrates from a 20 mg/ml toluene solution, were measured. The PL spectrum, peaked at 538 nm, was measured with an excitation at a wavelength of 379 nm, where the absorbance of the film was maximal (see Fig. 8.3). The solid-state PLQY was measured to be 61 $\pm$ 2 % with the same excitation wavelength, using the Hamamatsu Photonics C9920-02 system. A 5–30 mg/ml solutions were used to produce a range of film-thicknesses varying from 30 to 400 nm to study the thickness dependent sensing.

The fluorescence decay lifetime of a CDTG copolymer film was first measured with a conventional time-correlated single photon counting (TCSPC) system, illustrated in Fig. 8.4. The optical excitation was provided by a pulsed nitride LED (Picoquant PLS-8-2-066) with central emission wavelength of 457 nm in this work. The LED pulses, triggered by a signal generator at a repetition rate of

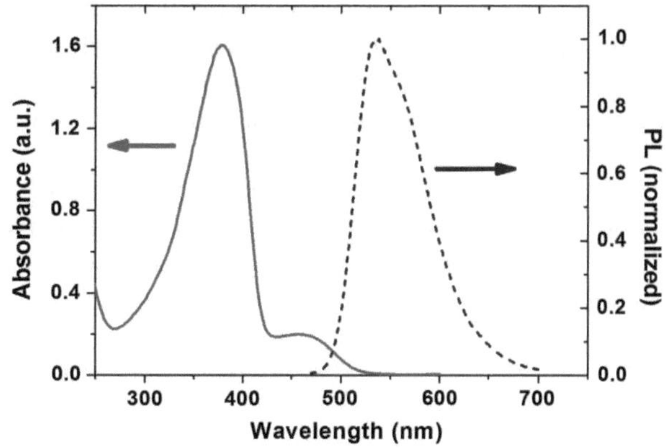

Fig. 8.3 Absorption (*solid line*) and PL spectra (*dashed line*) of a 250 nm thick film of CDTG copolymer

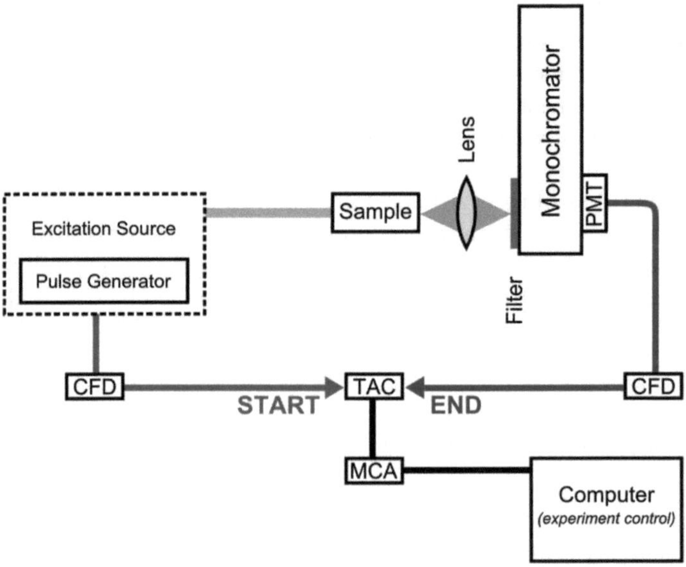

Fig. 8.4 Experimental set-up configuration of a conventional TCSPC measurement

100 kHz, had an approximate pulse duration of 1 ns FWHM. The signal generator also provided a "START" electronic pulse for the timing, passing through a constant fraction discriminator (CFD) to a time-to-amplitude converter (TAC). The CFD was used to minimise variations in the trigger timings. The TAC began charging when the "START" pulse arrived.

On the other side, the light emitted from the polymer film was collected by a lens and passed through a monochromator, to a micro-channel plate photomultiplier tube (MCP-PMT, Hamamatsu RU-3809U-50). The monochromator slit was set to be at the central wavelength of the detection, in this case 540 nm. The MCT-PMT converted the emitted photon energy into an electrical pulse, i.e. "STOP" pulse. It passed through a CFD to the TAC, stopping the charging of the TAC so that the period of time was determined. The TAC was then discharged to allow the multichannel analyser (MCA) to convert the electrical signal into a digital signal, i.e. a single count of detected photons during the certain time period. The count was then plotted on a histogram against time and the decay statistics were thus calculated.

The instrument response function (IRF) was measured with a light scattering colloidal silica, Ludox solution. The FWHM of the IRF of this apparatus was found to be 1.1 ns. The time-resolved data were fitted using iterative re-convolution to a bi-exponential decay:

$$y(t) = A_1 \exp(-\frac{t}{\tau_1}) + A_2 \exp(-\frac{t}{\tau_2}) \tag{8.1}$$

where A_1 and A_2 are pre-exponential factors. An average fluorescence lifetime is defined by:

$$\tau_{avg} = (A_1\tau_1 + A_2\tau_2)/(A_1 + A_2) \quad (8.2)$$

The non-linear fittings were performed using decay-analysis software FMAS DAS6 (IBH Ltd). A steady-state PL lifetime with a fast component of 1.3 ± 0.5 ns ($A_1 = 0.80$) and a slow component of 3.5 ± 0.2 ns ($A_2 = 0.20$) was measured with the CDTG film. The fast component was less accurate due to being close to the FWHM of IRF. The average of the PL lifetime was thus calculated to be 1.7 ns using Eq. 8.2.

8.2.3 Sensor Based on PL

The green fluorescence from a 50 nm thick CDTG film, optically excited by a 355 nm pulsed Nd:YVO_4 laser at 500 Hz repetition rate with an average power of 80 μW, responded rapidly to the 10 ppb DNB vapour, as shown in Fig. 8.5. Quenching efficiency was found to be 37 % (*ΔPL*) after a continuous 15-min DNB exposure.

The dynamics of the PL quenching as well as the recovery with a clean nitrogen purge, with similar flow-rate to that used in the DNB exposure, are shown in Fig. 8.6a, b respectively. The fluorescence intensity from the sample fully recovered back to its original level in nitrogen during the same period of time as DNB exposure. The cycle of exposure and recovery was repeated, as shown in Fig. 8.6c, d. This suggests that the CDTG sensor is capable for re-use in multiple explosive detections.

To investigate how the quenching depends on film thickness, a 50 nm and a 400 nm CDTG film were exposed to the same DNB exposure conditions and their PL quenching dynamics were recorded. Figure 8.7 compares the decreases in the PL intensity as a function of time between these two films. Although the absolute PL intensity from the thin film was much lower than the thick film, the thin film was quenched much faster in the first 10 min. The faster response of the thin film can be explained by the penetration speed of DNB molecules into the

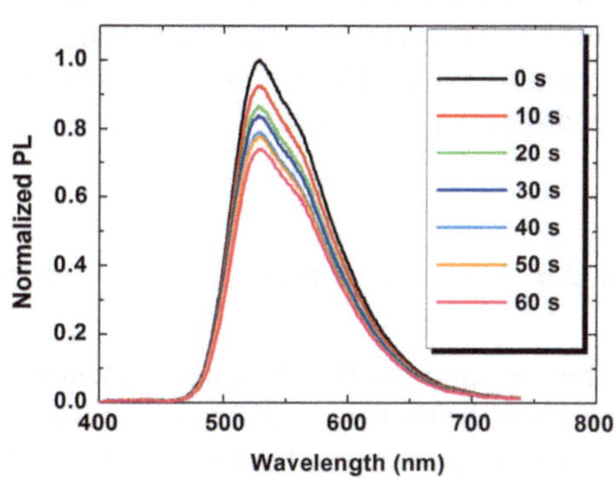

Fig. 8.5 Steady-state fluorescence spectra (normalized to the highest intensity) of a 50 nm CDTG film, measured at 0, 10, 20, 30, 40, 50, 60 s (*top to bottom*) after exposing to 10 ppb DNB *vapour* (ref. [5])

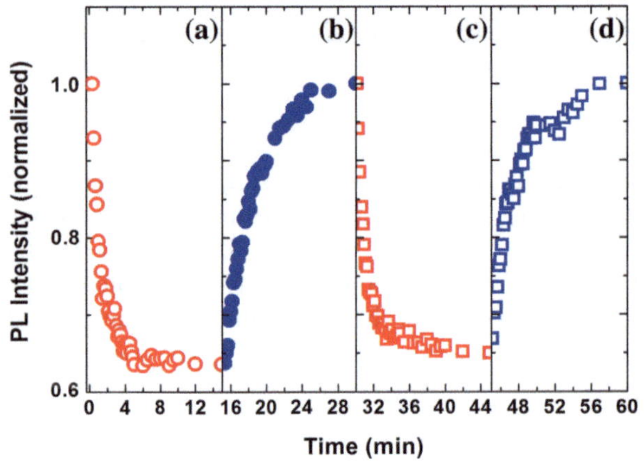

Fig. 8.6 Normalized PL dynamics monitored from a 50 nm CDTG film, during: **a** 1st 15-min exposure, **b** 1st 15-min recovery, **c** 2nd 15-min exposure and **d** 2nd 15-min recovery

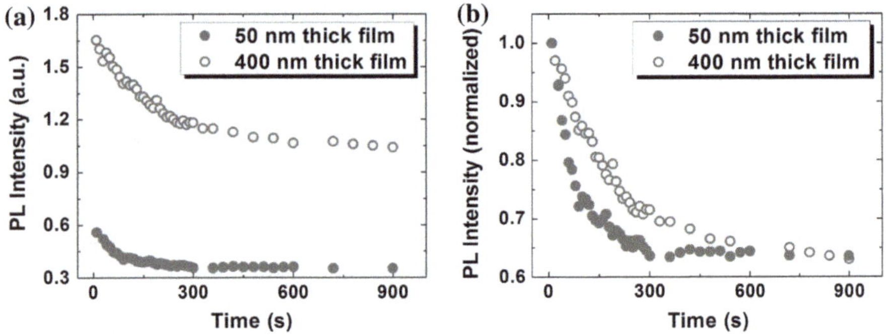

Fig. 8.7 Changes of PL intensity **a** un-normalized **b** normalized, from a 50 nm (*closed circle*) and a 400 nm (*open circle*) film respectively, during a 15-min exposure

polymer film. The DNB molecules diffuse into both films at a similar rate, and hence quench a larger fraction of the light emission from the thinner film. However, after a long exposure time, the quenching efficiencies of these two films were similar. This suggests that in real-time trace explosive detection, thinner films would be preferable due to their faster response to the limited amount of analyte, regardless of their weaker emission compared to thicker films.

8.2.4 Sensing Based on Conventional TCSPC System

The conventional TCSPC measurement system, described in Sect. 8.2.2, was used to measure the fluorescence lifetimes before and after a 15-min DNB exposure. Clean nitrogen was then used to recover the fluorescence lifetime in a further 15 min.

Table 8.1 PL lifetime measured with a conventional TCSPC before and after a 15-min exposure to DNB; Pre-exponential factors are shown in brackets; as well as the change of average lifetime in percentage ($\Delta\tau_{avg}$)

Sensor	Thickness (nm)	τ_{before}	τ_{after}	$\Delta\tau avg(\%)$
#1	50	1.3 ns (0.80)	0.97 ns (0.89)	32
		3.4 ns (0.20)	2.9 ns (0.11)	
#2	95	1.3 ns (0.84)	1.0 ns (0.89)	30
		3.5 ns (0.16)	3.0 ns (0.11)	
#3	150	1.3 ns (0.76)	0.91 ns (0.90)	38
		3.3 ns (0.24)	2.9 ns (0.10)	

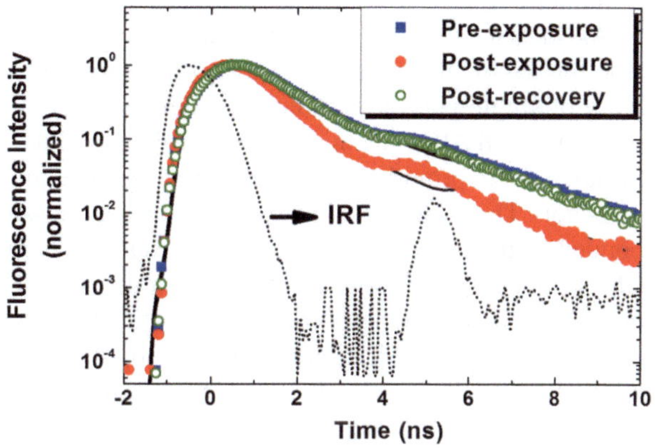

Fig. 8.8 Conventional TCSPC sensing with a CDTG sensor (thickness of 150 nm); *solid lines* are the bi-exponential fittings to the decays; *dotted line* is the instrument response function

Figure 8.8 shows the time-correlated PL decay curves from a 150 nm thick CDTG film measured with the TCSPC system; quenching of PL was observed after the film was exposed to the DNB molecules. The lifetime values in Table 8.1 were extracted from the decay curves by the fitting of bi-exponential expressions de-convolved with the IRF individually. After a 15- min DNB exposure, the average lifetime of the 150 nm thick sensor decreased from 1.78 ns to 1.11 ns, dropping by 38 % ($\Delta\tau_{avg}$), in good agreement with the decrease in the PL intensity for the same sample ($\Delta PL = 36$ %). After exposure, the PL was fully recovered by flowing clean nitrogen through the optical chamber, as shown in Fig. 8.8.

8.2.5 CMOS Lifetime Analysis Micro-System

The results in the last section have demonstrated successful sensing of 10 ppb explosive vapour based on measuring the PL lifetime of polymer films. As the lifetime value for a chosen material is relatively constant, i.e. not variable with film thickness, excitation intensity (not too high), collection efficiency or stray light in comparison with steady-state PL intensity, it can be considered a very practical sensing parameter. However, the conventional TCSPC system is bulky and expensive for on-site explosive detection.

In response to this Prof. Robert Henderson and Dr. Bruce Rae at the University of Edinburgh built a CMOS-based micro-system for time-resolved fluorescence lifetime measurements as part of the HYPIX project [7]. This micro-system has previously demonstrated lifetime measurements of colloidal quantum dots and Rhodamine dyes in solution [8]. The micro-TCSPC system combines a CMOS-driven micro-LED array excitation source, an array of photodetectors and lifetime signal processing circuitry, providing a compact, potentially low-cost approach for measuring the fluorescence decay lifetime of polymers. The work presented in this section was performed together with Dr. Bruce Rae from the University of Edinburgh.

As shown in Fig. 8.9, the polymer film was sandwiched between a Si/CMOS chip for driving the micro-LED array on the top and another Si/CMOS chip with an array of single photon avalanche diodes (SPAD) on the bottom. In this work, we used an 8 × 8 AlInGaN micro-LED array, fabricated at the IOP at the University of Strathclyde, to provide the optical excitation. The micro-LED pixels were 72 μm in diameter. The peak emitting wavelength was at 450 nm. The micro-LEDs were bonded with the Si/CMOS backplane using a bump-bonding process [9]. Each CMOS bonding area was connected to a micro-LED pixel by a gold bump-bond, providing each CMOS-controlled micro-LED pixel within a 200 μm pitch.

To measure the PL lifetime of the copolymer CDTG, the micro-LED pulse width was carefully set to 1.2 ns (FWHM), to achieve a reasonable signal-to-noise ratio of the detected signal, by adjusting the gate voltage of a current-starving transistor, i.e. controlling the charging and discharging time [8]. The pulse width of the micro-LED is critical for the resolution of the TCSPC system, as it directly affects the IRF and the excitation pulse intensity. The LED driving circuit was capable of producing a square-wave signal with a frequency range from 7 MHz to 800 MHz with the shortest micro-LED pulse width approximately 800 ps. The bottom CMOS chip includes a fully addressable 16 × 4 array of SPAD detectors

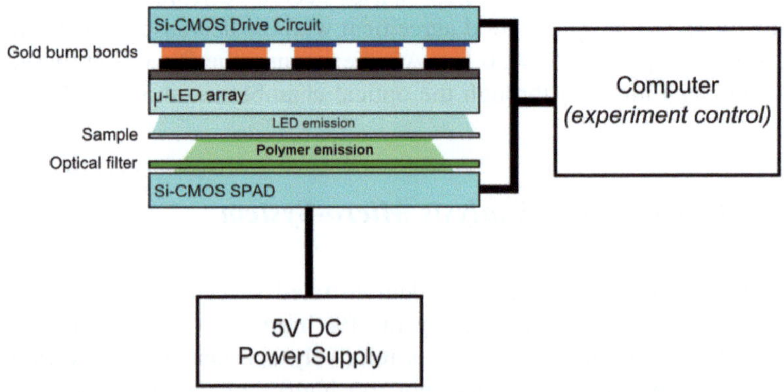

Fig. 8.9 Experimental set-up configuration of CMOS micro-TCSPC system

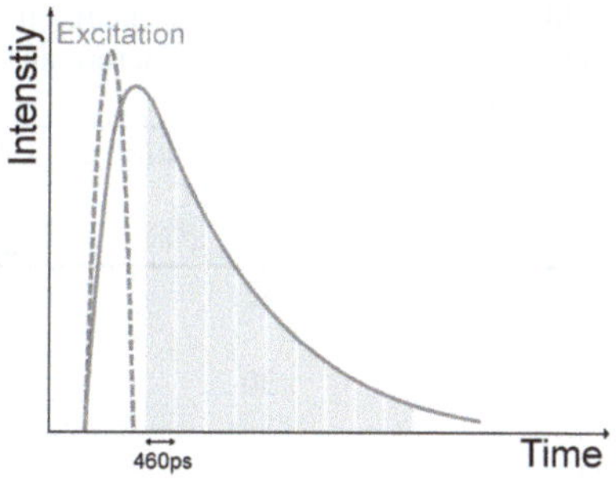

Fig. 8.10 Illustration of a fluorescence lifetime decay histogram measured with the micro-TCSPC

(aligned to the excitation source) integrated with on-chip timing and signal processing circuits. A 514 nm long pass filter (LP02-514RU-25, Semrock) was inserted between the micro-LEDs and the SPAD chips to block the residual micro-LEDs emission.

With a fresh CDTG film (take the 150 nm thick film as an example) spin-coated on glass substrate, placed between these two CMOS chips, the lifetime decay curve was captured within a series of time-gated windows. In other words, the emitted photons were "counted" in each time gate and the counts were then used to build a histogram of the decay curve, illuminated in Fig. 8.10. The time-gate width was set to be 460 ps, in order to cover a time window of ~20 ns in total. After exposing the CDTG film to 10 ppb DNB vapour for 15 min, the time-resolved decay curves of the film were re-measured, followed by a 15 min recovery in nitrogen. The decay curves were displayed and recorded with a laptop, and are presented in Fig. 8.11.

The solid lines in Fig. 8.11 are the bi-exponential fits post-processed with the recorded histograms. The average lifetime was calculated to be 2.66 ns with the clean sample and 1.63 ns after DNB exposure, a reduction of 38 %. The percentage changes in lifetime, $\Delta\tau_{avg}$ presented in Table 8.2, are consistent with the results measured with the conventional TCSPC. It is worth mentioning that the

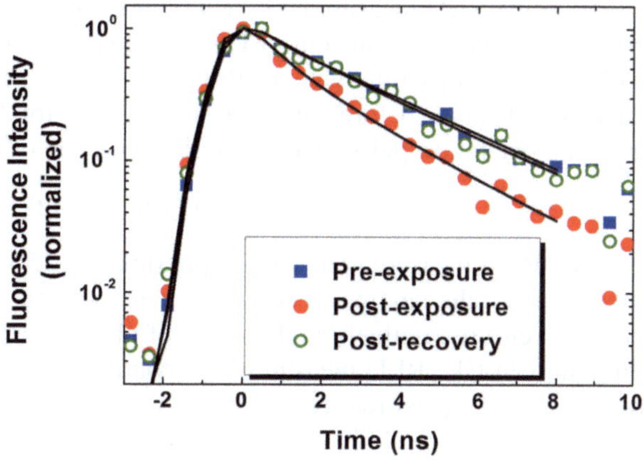

Fig. 8.11 PL decay measured using the CMOS micro-TCSPC system with a 150 nm thick CDTG film before (*blue squares*), after DNB exposure (*red circles*), and after recovery in nitrogen (*green open circles*)

Table 8.2 Comparison of the change in lifetime measured with both the conventional TCSPC and the CMOS micro-system, before and after a 15-min exposure to DNB with pre-exponential factor shown in bracket

Sensor	Thickness (nm)	Conventional TCSPC $\Delta\tau_{avg}(\%)$	CMOS Micro-TCSPC τ_{before}	τ_{after}	$\Delta\tau_{avg}(\%)$
#1	50	32	0.55 ns (0.30) 3.3 ns (0.70)	0.36 ns (0.53) 2.8 ns (0.47)	39
#2	95	30	0.51 ns (0.34) 3.4 ns (0.66)	0.40 ns (0.58) 2.7 ns (0.42)	43
#3	150	38	0.68 ns (0.24) 3.3 ns (0.76)	0.53 ns (0.50) 2.7 ns (0.50)	38

fast components have low accuracy the FWHM of the IRF is approximately 1.3 ± 0.23 ns.

8.2.6 Ultra-Portable Organic/Inorganic Explosives Sensor

The CMOS/polymer explosive sensor based on the micro-TCSPC system has shown great potential to be used in field as it is small in size, easy to operate, could be potentially powered by battery and also wirelessly communicate with a laptop. However, with the configuration in Sect. 8.2.5 the lifetime value was not available without post-data-processing. A rapid lifetime determination (RLD) method was developed by the collaborators at the University of Edinburgh, in which the PL decay of the polymer film was approximated to be a single-exponential decay, the lifetime τ of which is determined by:

$$I(t) = I_0 \exp(-\frac{t}{\tau}) \tag{8.3}$$

where $I(t)$ is fluorescence intensity at time t, I_0 is initial fluorescence intensity, i.e. $t = 0$. In the RLD measurements with the CMOS micro-system, the accumulated counts (A1 and A2 in Fig. 8.12), detected in two equal-time windows ΔT ($\Delta T = 4 \times 460$ ps time-gate in this work), were recorded, taking into account background counts beforehand. The starting time for A1 was set to be immediately after the excitation pulse. The decay lifetime τ is calculated using Eq. 8.4.

$$\tau = \Delta T / \ln(A1/A2) \tag{8.4}$$

Each RLD measurement took approximately 15 s to display a calculated lifetime value on the laptop screen, with the integration time for each time gate set to be 4 s to ensure relatively high counts of photons detected. The laptop-controlled program can take RLD measurements continuously, allowing lifetime changes to be observed in quasi-real-time.

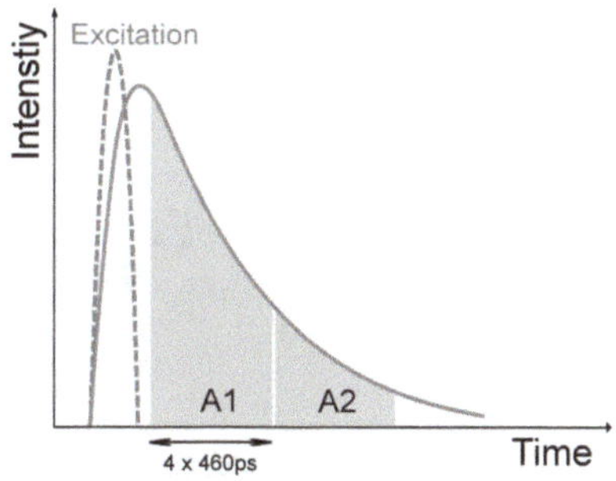

Fig. 8.12 Illustration of a two-gate rapid-lifetime determination of fluorescence lifetime

To simulate an on-site explosive detection, the micro-TCSPC system with CDTG films were all placed in a customized sensing box of dimensions 20 × 13 × 7 cm, as shown in Fig. 8.13. Several CDTG films with different thicknesses were exposed to 10 ppb DNB vapour one at a time, allowing the study of the effect of film thickness. Following exposure, the gas port of the box was switched to clean nitrogen allowing the kinetics of the lifetime recovery to be monitored. The lifetime of the CDTG film was calculated on-chip in every 15 s for 15 min continuously. Taking the 150 nm sensor for an example, Fig. 8.14 shows the dynamics of the PL lifetime in a 15-min exposure time. The lifetime decreased from 2.95 ns to 1.93 ns and after 15-min recovery in nitrogen increased back to 2.93 ns. The sensing efficiency ($\Delta\tau$) of this sensor was approximately 35 %. It should be noted that as the rapid-lifetime determination method assumed the fluorescence decay was a single-exponential curve, the CDTG lifetime values were slightly over-estimated in the RLD mode, but this was deemed within an acceptable range as the longer lifetime component of the decay contributes more to the overall calculation of the average lifetime.

The RTD results with different thickness films are summarized in Table 8.3. $\Delta\tau$ values were found to be independent of the variation of the thickness of the CDTG films, suggesting that the DNB molecules could penetrate into both thin and thick CDTG films in the 15-min period of the measurement time.

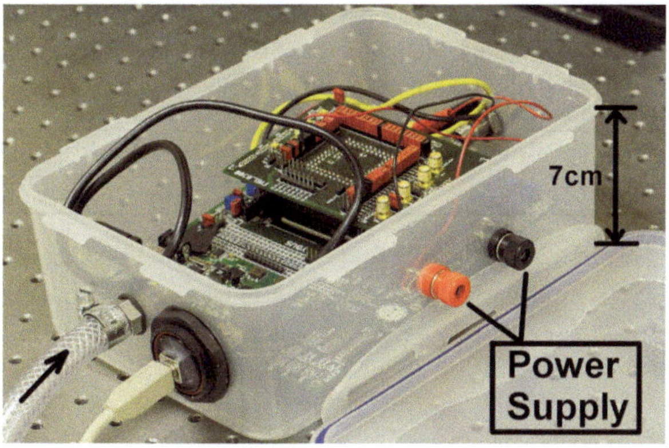

Fig. 8.13 A photograph of the customized sensing box, consisting of two gas ports, two wires for connection to an external 5 V DC power supply and a USB connection for data communication with a laptop, the CMOS TCSPC system can be seen sitting inside

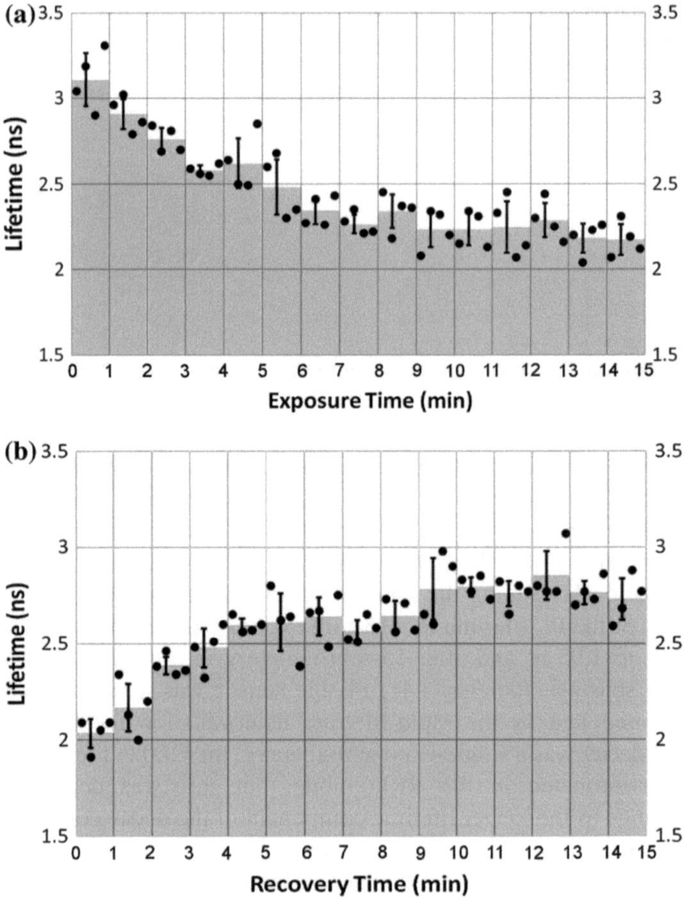

Fig. 8.14 Real-time two gate RLD dynamics during **a** 15-min DNB exposure, and **b** 15-min nitrogen recovery, of a 150 nm sensor; the scatter points (*closed circles*) represent the lifetime values calculated every 15 s, and the background green bars represent the average lifetime in every minute with standard deviation error bars (*solid line*)

Table 8.3 In-situ CMOS lifetime of CDTG samples before ($\tau_{pre\text{-}exp}$) and after a 15-min exposure to 10 ppb DNB ($\tau_{post\text{-}exp}$), and after a 15-min recovery ($\tau_{post\text{-}rec}$), as well as the change of lifetime ($\Delta\tau$) during the exposure time

Sensor	Film thickness(nm)	CMOS real-time TCSPC lifetime			
		$\tau_{pre\text{-}exp}$(ns)	$\tau_{post\text{-}exp}$(ns)	$\tau_{post\text{-}rec}$(ns)	$\Delta\tau$(%)
#1	50	2.85	1.70	2.90	40
#2	95	2.82	1.80	2.90	36
#3	150	2.95	1.93	2.95	35
#4	200	3.00	1.92	2.94	36
#5	250	3.00	2.05	2.90	32
#6	330	2.95	1.95	2.72	34
#7	400	3.20	2.00	2.73	38

8.3 Hybrid Organic Laser Sensor

The CMOS/GaN/polymer sensor was a big step towards an ultra-portable trace explosive sensing device, with on-chip near-real-time data processing and potentially remote control. As discussed in Chap. 7, explosive sensors based on polymer lasers have much faster responsivity compared to sensors based on PL. In this last section, a low-cost LED pumped polymer laser sensor is investigated.

To make such a sensor, the polymer lasers are required to have a threshold low enough to be pumped by an LED far beyond the lasing threshold and therefore to be able to show changes in the laser emission after exposure to the target analyte. The low threshold polymer was chosen to be BBEHP-PPV, studied in detail in previous chapters. As mentioned in Chap. 7, the same polymer has been used to make a microring laser which has shown sensing ability to TNT and DNT vapours [10]. To confirm its potential for sensing of explosives, PL sensing measurements using BBEHP-PPV thin films were performed first. After 90 s continuous exposure, the PL intensity of an 80 nm thick film decreased by 15 %, as shown in Fig. 8.15. Recovery of the PL emission from the exposed film was achieved with a longer time period of clean-nitrogen purge.

An InGaN LED pumped BBEHP-PPV laser sensor was next demonstrated. Fig. 8.16 shows the laser output as a function of LED peak intensity before and after exposure to 10 ppb DNB vapour for only 90 s. The lasing threshold increased by 38 %, from 16 A/pulse (corresponding to 535 W/cm^2) to 22 A/pulse (corresponding to 661 W/cm^2), following a 90- s DNB exposure. The laser emission output dropped by 62 %, at a fixed driving current of 38 A/pulse, during the 90 s exposure time.

Although in this work a fibre coupled CCD spectrograph was used to record the polymer laser spectra, this LED pumped polymer laser sensor could be easily integrated with a compact photodetector into a portable sensing device, with high sensitivity and fast responsivity.

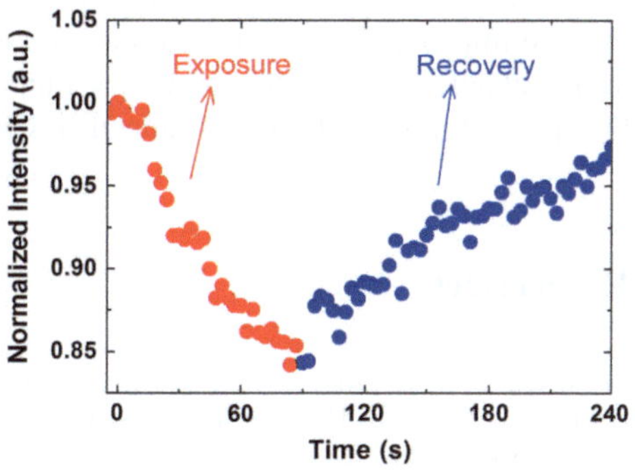

Fig. 8.15 Sensing dynamics based on PL from an 80 nm film of BBEHP-PPV

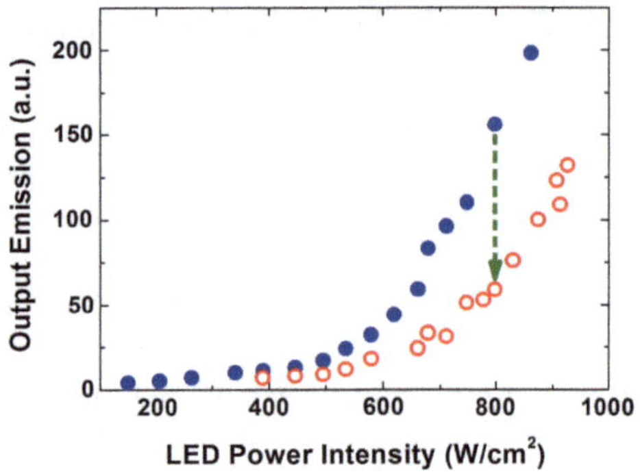

Fig. 8.16 Power characteristics of an InGaN LED pumped BBEHP-PPV laser before and after a 90-sec exposure to the model explosive, 10 ppb 1,4-DNB

8.4 Conclusions

We have shown that the conjugated polymer CDTG, in combination with a CMOS PL lifetime detection system, can be used as a sensor for detecting low vapour pressure nitro-aromatic molecules. For both conventional and CMOS-based time-resolved fluorescence lifetime system, successful indication of the presence of the analyte has been achieved. Compared to the conventional TCSPC, the CMOS micro-system provides both a simple and reliable technique for PL lifetime measurement, with potential for practical applications such as in-field explosive detection. This is the first demonstration of integration of CMOS/gallium nitride/organic semiconductors. At the end of this chapter, an LED pumped polymer laser is also used as an explosive sensor providing very fast responsivity. With the possibility of a CMOS driven LED pumped polymer laser, the laser sensor has potential of being even more integrated and intelligent.

From this chapter, we can see the potential for inexpensive, ultra-portable and sensitive explosive detectors using hybrid systems of inorganic and organic semiconductors. One potential area of application is to clear landmines left after wars. Humanitarian demining is a challenging problem and polymer laser sensors could be used to complement existing technologies such as ground-penetrating radar to improve the detection of hazardous objects. Aspects of the work presented in this chapter will be developed for humanitarian demining within the European FP7-project TIRAMISU, introduced in Chap. 9.

References

1. Woodfin, R. L. (2007). *Trace chemical sensing of explosives*. Newyork: Wiley-Interscience.
2. Fido® Explosives Detector Technical Review- online document. http://gs.flir.com/detection/explosives/fido/
3. Cavaye, H., Shaw, P. E., Wang, X., Burn, P. L., Lo, S. C., Meredith, P. (2010). Effect of dimensionality in dendrimeric and polymeric fluorescent materials for detecting explosives. *Macromolecules*, *43* (24), 10253–10261.

4. Olley, D. A., Wren, E. J., Vamvounis, G., Fernee, M. J., Wang, X., Burn, P. L., Meredith, P., Shaw, P. E. (2011). Explosive sensing with fluorescent dendrimers: The role of collisional quenching. J. Mater. Chem, *23* (3), 789–794.
5. Wang, Y., Rae, B. R., Henderson, R. K., Gong, Z., Mckendry, J., Gu, E. D., et al. (2011). Ultra-portable explosives sensor based on a CMOS fluorescence lifetime analysis micro-system. *Journal – AIP Advances, 1*(3), 032115.
6. Wang, Y., Turnbull, G. A., Samuel, I. D. W. (2011). Conjugated polymer sensors for explosive vapor detection. *Proceedings of SPIE, 8118*(1).
7. Rae, B. R., Muir, K. R., Gong, Z., McKendry, J., Girkin, J. M., Gu, E., et al. (2009). A CMOS Time-resolved fluorescence lifetime analysis micro-system. *Sensors-Basel, 9*(11), 9255–9274.
8. Rae, B. R., Yang, J. B., McKendry, J., Gong, Z., Renshaw, D., Girkin, J. M., et al. (2010). A vertically Integrated CMOS microsystem for time-resolved fluorescence analysis. *IEEE Transactions on Biomedical Circuits and Systems, 4*(6), 437–444.
9. McKendry, J. J. D., Rae, B. R., Gong, Z., Muir, K. R., Guilhabert, B., Massoubre, D., et al. (2009). Individually addressable alingan micro-led arrays with CMOS control and subnanosecond output pulses. *IEEE Photonics Technology Letters, 21*(9–12), 811–813.
10. Rose, A., Zhu, Z. G., Madigan, C. F., Swager, T. M., & Bulovic, V. (2005). Sensitivity gains in chemosensing by lasing action in organic polymers. *Nature, 434*(7035), 876–879.

Chapter 9
Conclusions and Perspectives

9.1 Conclusions

This thesis has focused on lasers and chemosensors fabricated using organic semiconductors. Organic semiconductors are novel materials made from conjugated molecules. One of their main advantages is that they are easily processed from solutions. This means that the final device can be printed onto a number of different substrates, including for example flexible and transparent plastics. Organic semiconductors also benefit from the ability to synthesise different chemical structures, giving colour-tunable luminescence across the whole visible spectrum. After two decades of development of both the chemistry and device fabrication, various products have been realised such as OLED displays and lighting.

In organic semiconductor lasers, amplification is provided by the organic molecules. A good gain material exhibits both low self-absorption (i.e. low loss) and high gain coefficients. The optical cavities used in this thesis are all based on the distributed feedback (DFB) resonators. To make a surface emitting DFB laser, we spin coat a thin layer of conjugated polymer onto a periodically corrugated glass substrate to form an air-polymer-silica waveguide where the first order diffraction provides output coupling and the second order diffraction contributes to the in-plane feedback. The lasing wavelength is then determined by the Bragg condition which includes the effective refractive index of the waveguide and the period of the corrugation [1].

In the first experimental chapter, Chap. 4, novel star-shaped gain materials, oligofluorene truxene T3 and T4, were investigated. Photophysical studies highlighted the key features of these materials such as their high solid-state photoluminescence quantum yield and low optical losses that lead to very low amplified spontaneous emission thresholds. The materials were shown, by ellipsometry measurement, to form isotropic thin films, a fact that contributed to their low waveguide loss coefficient. All of the above properties demonstrated that these materials were promising for application in organic semiconductor lasers. Surface emitting DFB lasers made with these oligofluorene truxenes have shown very low

Y. Wang, *Low Threshold Organic Semiconductor Lasers*, Springer Theses,
DOI: 10.1007/978-3-319-01267-4_9, © Springer International Publishing Switzerland 2014

lasing threshold, 270 W/cm^2, and the broadest tunability, over 50 nm, in the blue spectral region for organic semiconductor lasers.

The truxene lasers in Chap. 4 were optically pumped by a Q-switched microchip laser. To date electrically driven lasing in organic semiconductors is still an outstanding challenge. Indirect electrical pumping using an electrically controlled light-emitting device as an interface is one solution to this challenge. This has been investigated in Chaps. 5 and 6.

In 2008, an InGaN light emitting diode (LED) was first used to pump a copolymer ADS233YE laser just above threshold [2]. In the last few years, significant developments have occurred in the field of inorganic LEDs as well as synthetic organic semiconductors. Very low threshold surface-emitting BBEHP-PPV lasers were fabricated using one-dimensional DFB resonators in Chap. 5. The lowest threshold achieved with 4 ns OPO pulse pumping was 118 W/cm^2 with a lasing peak-wavelength at 533 nm. A high power, inexpensive Philips LUMILED LED with 448 nm peak emission was selected to pump BBEHP-PPV polymer lasers in Chap. 5. A maximum pulsed intensity of 1,000 W/cm^2 is available from this LED driven at 20 Hz, with a light-emitting area of 1.3 $\times$ 1.3 mm. Different from the OPO, the LED pulse width is a few 10 s ns, with a slower rise time. When pumped by the LED, a narrow-linewidth BBEHP-PPV laser peak appears at 533 nm for drive currents above 15.7 A/pulse, corresponding to 500 W/cm^2. Its higher pump threshold, in comparison with using the OPO as the pump source, is mainly due to the long LED pulse and slow rising edge. This demonstration is the first time that an inorganic LED has been used to pump an organic laser far above threshold, allowing these hybrid lasers' output beam divergence, as well as pulse duration, to be studied in detail. An interesting 'double-threshold' phenomenon was also observed and explained in terms of the long-pulse pumping condition.

The HYPIX project [3] is aiming to fully hybridise organic semiconductor lasers with CMOS based electronic control circuitry by using intermediate gallium nitride (GaN) micro-pixellated LEDs. In Chap. 6, mixed-order DFB resonators were fabricated by UV-nanoimprint lithography (NIL). These resonators were designed to limit the output coupling efficiency in order to further reduce the polymer laser threshold. Nanoimprinted gratings, with feature sizes of sub-hundred nm, have been used to fabricate BBEHP-PPV polymer lasers. Different pump beam geometries have been investigated to optimize the laser threshold. Very low polymer laser thresholds of approximately 50 W/cm^2 using the OPO as the pump source have been achieved with the NIL mix-order gratings. This threshold is more than an order of magnitude lower than any reported NIL organic semiconductor lasers. UV-NIL polymer lasers have been successfully pumped by customized micro-LED arrays, with thresholds of around 250 W/cm^2. Such hybrid lasers present a highly compact and low-cost laser source. The compatibility of micro-LEDs with CMOS based control, brings hope of hybridized organic/inorganic smart pixel arrays for promising applications, including explosive sensing has been discussed in Chaps. 7 and 8.

Sensing using organic semiconductor films is based on the fact that the presence of explosive molecules alters the film's fluorescence characteristics. To achieve highly sensitive and fast response polymer based explosive sensors, the morphology of the polymer is one of the important factors due to its direct effect on the penetration of the explosive molecules into the film. In Chap. 7, a polymer of intrinsic micro-porosity (PIM-1) was introduced. Fast indication of the presence of the analyte, 10 ppb 1,4-DNB, by monitoring the photoluminescence was achieved. By using a DFB laser geometry, a significant enhancement in sensing efficiency and responsivity has been demonstrated. The response accumulates, suggesting a very strong interaction between the DNB molecules and the microporous polymer, leads to an inherent memory ability. The combination of ease of processing into thin coatings, its intense fluorescence and its microporous structure makes PIM-1 a very promising material for use in low-cost rapid sensing of ultra-low-concentration explosives.

The final experimental chapter, Chap. 8, explored ways to make the polymer sensors more portable. A CMOS micro-system provided both a simple and reliable technique for PL lifetime measurement, which has shown capability in trace explosive detection. The integration of CMOS/gallium nitride/organic semiconductors demonstrates a potentially inexpensive and portable explosive detector. Furthermore, an LED pumped BBEHP-PPV laser has also been used to detect low vapour pressure explosives with very fast responsivity. With the possibility of a CMOS driven micro-LED pumped polymer laser, the hybrid laser sensor has the potential of being even more integrated and intelligent. We can see the prospect of a robot carrying these high performance sensors into dangerous areas for remote demining.

9.2 Outlook

The sensing work presented in this thesis has been highlighted in public media, including LaserFocusWorld [4], ScienceDaily [5] and Chemistry World [6]. This recognition illustrates the impact of the work on organic semiconductor explosive sensors.

The organic sensors can detect vapours that are strongly electron-deficient. In minefield type conditions, this type of sensors can offer low false alarm rates, however, for more complex environments such as airports, identification of detected explosive is required. To achieve this goal, our organic sensors need to be selective. There is also potential to improve the lower limit of detectable concentrations, i.e. the device's sensitivity.

To address these points one possible solution I propose is to integrate multiple functionalized organic semiconductor lasers on to a single chip with millimetre dimensions. This would open up the possibility of differentiating between different explosive materials, all with a single optical pump source. By using an array of lasers made from different materials (with distinct responsivity to various types of explosives) [7] and diverse laser cavities (providing variable sensitivity), an "organic laser nose" could be fabricated, see Fig. 9.1, and so giving a step-change in plastic sensor's performance and capability.

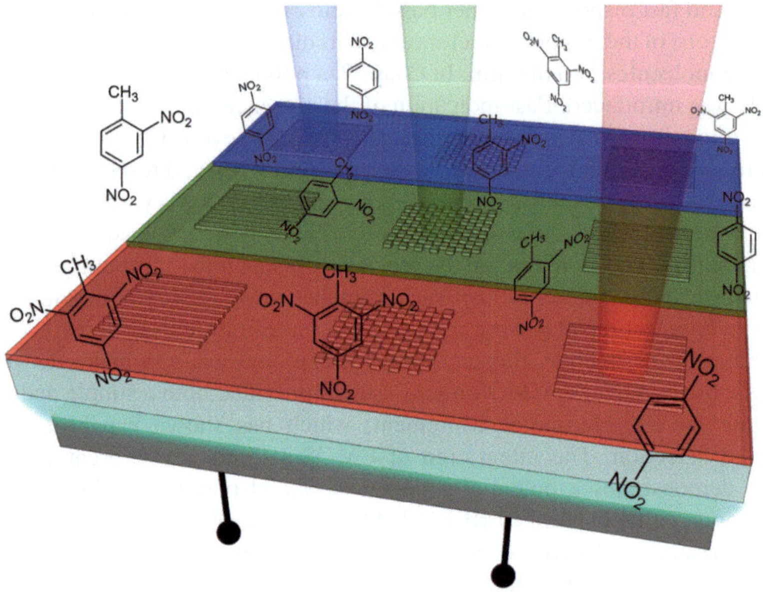

Fig. 9.1 Schematic image of a multi-pixel "organic laser nose"

Leading from the initial results in this thesis, a European project TIRAMISU (standing for toolbox implementation for removal of anti-personal mines, submunitions and UXOs (Unexploded Ordnances)) has been awarded, with St Andrews as one of the principal partners [8]. The objective of the TIRAMISU project is to provide the foundation for a global toolbox that will cover the main mine action activities related to Humanitarian Demining, from the survey of large areas to the actual disposal of explosive hazards. I will continue my work on organic explosive sensors as part of this project.

Finally, although using the CMOS driven micro-LED arrays to pump organic lasers has not been demonstrated yet, due to the limited pulse current the CMOS driver could provide, I believe that the CMOS controlled hybrid laser array could be realized in the near future. This would lead to an innovative laser source and a very portable smart explosive detection device.

References

1. Samuel, I. D. W., & Turnbull, G. A. (2007). Organic semiconductor lasers. *Chemical Reviews, 107*(4), 1272–1295.
2. Yang, Y., Turnbull, G. A., Samuel, I. D. W. (2008). Hybrid optoelectronics: A polymer laser pumped by a nitride light-emitting diode. *Applied Physics Letters, 92*(16), 163306.
3. HYPIX project. http://hypix.photonics.ac.uk/.

4. Overton, G. (2011). Fluorescence sensors lunchbox-sized fluorescence sensor detects explosive vapors. *Laser Focus World, 47*(12), 18.
5. ScienceDaily-New device exposes explosive vapors. http://www.sciencedaily.com/releases/2011/08/110815113609.htm.
6. Davies, E. (2012). *Sniffing out explosives* (pp. 48–52). World May: Chemistry.
7. Thomas, S. W., Joly, G. D., Swager, T. M. (2007). Chemical sensors based on amplifying fluorescent conjugated polymers. *Chemical Reviews, 107*, 1339–1386.
8. TIRAMISU project. http://www.fp7-tiramisu.eu/.

Curriculum Vitae

Education

09/2008–09/2012 **PhD** at University of St Andrews
09/2007–09/2008 **Masters** in "Photonics and Optoelectronic Devices"
University of St Andrews & Heriot-Watt University
Thesis: Laser-Based Photoacoustic Spectroscopy and Gas Detection
09/2003–09/2007 **Bachelors** in "Electronics Science & Technology (Optoelectronics)"
East China Normal University
Thesis: Integrated Detector and Interferometer for Near-Infrared Single Photons

Research Highlights & Prizes

- 2012—**EPSRC** Doctoral prize for improving retention of the very best students in research careers (one-year research grant)
- 2012—Best student presentation, symposium on Organic Photonics & Electronics at the **EOS (European Optical Society)** Annual Meeting
- 2012–Best poster, **SID (Society for Information Display)** Organic Electronics UK
- 2011–The **AIP Advances paper** on CMOS Time Resolved PL sensor featured in a news highlight and press release by American Institute of Physics (AIP)
- 2011–"Optoelectronics Committee Prize" by **The Rank Prize Funds** for "the Best Contributed Paper" at the symposium on Solid-State Lighting
- 2011–**School of Physics & Astronomy** 3rd year postgraduate students poster presentation prize

Y. Wang, *Low Threshold Organic Semiconductor Lasers*, Springer Theses,
DOI: 10.1007/978-3-319-01267-4, © Springer International Publishing Switzerland 2014

Research Interests

Organic semiconductors have great potential as low cost, easy to make, tuneable visible wavelength sources. My current research is to develop hybrid optoelectronics by combining organic semiconductor lasers with inorganic LEDs, and their applications in explosives detection. I have focused on the studies of photo-physical properties of a wide range of organic semiconductors used as gain materials for lasers and have developed feedback resonators and pump sources for organic semiconductor lasers. I am also developing methods for improving the sensitivity and selectivity of the organic semiconductor explosives sensors.

Conference Presentations

Vth International Krytyn Summer School, Krutyn, Poland, June 2009, poster presentation entitled "Low-threshold organic lasers based on oligofluorene truxenes"

UK Semiconductors, Sheffield, June 2010, oral presentation entitled "Advanced materials for low-threshold organic semiconductor lasers"

The Rank Prize Funds (Optoelectronics Symposium), Lake District, September 2011, oral presentation entitled "InGaN LED pumped polymer laser explosive sensor"

SPIE Photonics Europe conference, Belgium, April 2012, oral presentation entitled "InGaN LED pumped polymer laser explosive sensor"

EOS (European Optical Society) Annual Meeting, TOM 5 Organic Photonics & Electronics, Aberdeen, September 2012, oral presentation entitled "Nanoimprinted polymer lasers pumped by light-emitting diodes"

SID (Society for Information Display) Organic Electronics, London, September 2012, poster presentation entitled "Organic semiconductor explosive sensors"

Publications

1. J. Herrnsdorf, Y. Wang, J. J. D. McKendry, D. Massoubre, Z. Gong, B. Guilhabert, G. Tsiminis, G. A. Turnbull, I. D. W. Samuel, N. Laurand, E. Gu and Martin D. Dawson, "Micro-LED Pumped Polymer Laser: A Discussion of Future Pump Sources for Organic Lasers," *Laser & Photonics Reviews* Accepted for publication (2013)
2. Y. Wang, A. L. Kanibolotsky, P. J. Skabara, I. D. W. Samuel and G. A. Turnbull, "LED pumped polymer laser sensor for explosives," *Laser & Photonics Reviews* Accepted for publication (2013)
3. E. R. Martins,* Y. Wang,* A. L. Kanibolotsky, P. J. Skabara, I. D. W. Samuel and G. A. Turnbull, "Low-threshold nanoimprinted lasers using substructured gratings for control of distributed feedback," *Adv. Opt. Mater.* **1**, 563–566 (2013)
   * Joint lead author
4. Y. Wang, G. Tsiminis, A. L. Kanibolotsky, P. J. Skabara, I. D. W. Samuel and G. A. Turnbull, "Nanoimprinted polymer lasers with threshold below 100 W/cm^2 using mixed-order distributed feedback resonators," *Opt. Express* **21**, 14362–14367 (2013)
5. G. Tsiminis,* Y. Wang,* A. L. Kanibolotsky, P. J. Skabara, I. D. W. Samuel and G. A. Turnbull, "Nanoimprinted organic semiconductor laser pumped by a light-emitting diode," *Adv. Mater.* **25**, 2826-2830 (2013).
   * Joint lead author
6. Y. Wang, Y. Yang, G. A. Turnbull and I. D. W. Samuel, "Explosive Sensing Using Polymer Lasers," *Mol. Cryst. Liq. Cryst.* **554**, 103-110 (2012).
7. Y. Wang, B. R. Rae, R. K. Henderson, Z. Gong, J. Mckendry, E. Gu, M. D. Dawson, G. A. Turnbull and I. D. W. Samuel, "Novel Ultra-portable Explosives Sensor Based on a

CMOS Fluorescence Lifetime Analysis Micro-System," *AIP Adv.* **1** (3), 032115 (2011).

8. Y. Wang, G. A. Turnbull and I. D. W. Samuel, "Conjugated polymer sensors for explosive vapor detection," *Organic Semiconductors in Sensors and Bioelectronics IV, Proc. of SPIE* **8118**, (2011).

9. Y. Wang, N. B. McKeown, K. J. Msayib, G. A. Turnbull and I. D. W. Samuel, "Laser Chemosensor with Rapid Responsivity and Inherent Memory Based on a Polymer of Intrinsic Microporosity," *Sensors* **11** (3), 2478-2487 (2011).

10. Y. Chen, J. Herrnsdorf, B. Guilhabert, A. L. Kanibolotsky, A. R. Mackintosh, Y. Wang, R. A. Pethrick, E. Gu, G. A. Turnbull, P. J. Skabara, I. D. W. Samuel, N. Laurand, M. D. Dawson, "Laser action in a surface-structured free-standing membrane based on a pi-conjugated polymer-composite," *Org. Electron.* **12** (1), 62-69 (2011).

11. Y. Wang, G. Tsiminis, Y. Yang, A. Ruseckas, A. L. Kanibolotsky, I. F. Perepichka, P. J. Skabara, G. A. Turnbull and I. D. W. Samuel, "Broadly tunable deep blue laser based on a star-shaped oligofluorene truxene," *Synth. Met.* **160** (13-14), 1397-1400 (2010).

12. G. Tsiminis, Y. Wang, P. E. Shaw, A. L. Kanibolotsky, I. F. Perepichka, M. D. Dawson, P. J. Skabara, G. A. Turnbull and I. D. W. Samuel, "Low-threshold organic laser based on an oligofluorene truxene with low optical losses," *Appl. Phys. Lett.* **94** (24), 243304 (2009).

MIX
Papier aus verantwortungsvollen Quellen
Paper from responsible sources
FSC® C105338

If you have any concerns about our products,
you can contact us on
ProductSafety@springernature.com

In case Publisher is established outside the EU,
the EU authorized representative is:
Springer Nature Customer Service Center GmbH
Europaplatz 3, 69115 Heidelberg, Germany

Printed by Libri Plureos GmbH
in Hamburg, Germany